AF581717

Recent Advances in Mathematical Aspect in Engineering

Recent Advances in Mathematical Aspect in Engineering

Editors

Rahmat Ellahi
Sadiq M. Sait
Huijin Xu

MDPI • Basel • Beijing • Wuhan • Barcelona • Belgrade • Manchester • Tokyo • Cluj • Tianjin

Editors
Rahmat Ellahi
King Fahd University of Petroleum and Minerals
Dhahran
Saudi Arabia

Sadiq M. Sait
King Fahd University of Petroleum & Minerals
Saudi Arabia

Huijin Xu
Shanghai Jiao Tong Univeristy
China

Editorial Office
MDPI
St. Alban-Anlage 66
4052 Basel, Switzerland

This is a reprint of articles from the Special Issue published online in the open access journal *Symmetry* (ISSN 2073-8994) (available at: https://www.mdpi.com/journal/symmetry/special_issues/Recent_Advances_Mathematical_Aspect_Engineering).

For citation purposes, cite each article independently as indicated on the article page online and as indicated below:

LastName, A.A.; LastName, B.B.; LastName, C.C. Article Title. *Journal Name* **Year**, *Volume Number*, Page Range.

ISBN 978-3-0365-7490-5 (Hbk)
ISBN 978-3-0365-7491-2 (PDF)

Contents

About the Editors

Rahmat Ellahi

Professor Rahmat Ellahi, PhD, has achieved great academic standing from the University of Punjab, Quaid-i-Azam University, Islamabad, Pakistan, and the University of California Riverside, USA, as his alma mater. He has published around 300 papers in journals based in the USA, Germany, the UK, Canada, etc. His work has been cited more than 23,000 times on Google Scholar and has an h-index of 88. He is an editor/editorial board member of over 19 international journals. Additionally, he has edited seven Special Issues for ISI impact factor journals. He has successfully supervised 35 PHd/MS research students. His leadership in academics is further reflected through his collaboration with more than 85 international leading scientists from all over the world. He is an author of six books that have been published at the national and international levels and has organized eight international conferences. Besides having several other honors, he is the recipient of several national and international awards such as the Fulbright Award by the USA, Highly Cited Researcher Awards by Clarivate Analytic, Best University Teacher Award by HEC, Productive Scientist of Pakistan Awards by PCST, in Category A, Best Paper Award by Emerald Literati, Gold Medal from the Pakistan Academy of Sciences and Best Book Award, etc. He is topmost among Pakistani Scientists, being in the top 2% of highly cited scientists of the world as per Stanford University USA for the years 2020, 2021, and 2022. In keeping with his excellent research work, the United State of Education Foundation honored him as a Fulbright Fellow and the King Fahd University of Petroleum and Minerals, Dhahran, Saudi Arabia, honored him with the Chair Professor at Research Institute KFUPM.

Sadiq M. Sait

Sadiq M. Sait was born in Bengaluru. He received the bachelor's degree in electronics engineering from Bangalore University in 1981, and the master's and Ph.D. degrees in electrical engineering from the King Fahd University of Petroleum & Minerals (KFUPM) in 1983 and 1987, respectively. He is currently a Professor of Computer Engineering and the Director of the Office of Industry Collaboration, KFUPM. He has authored over 200 research papers, contributed chapters to technical books, granted several US patents, and lectured in over 25 countries. He is also the Principle Author of two books. He received the Best Electronic Engineer Award from the Indian Institute of Electrical Engineers, Bengaluru, in 1981.

Huijin Xu

Huijin Xu is now an associate professor in Shanghai Jiao Tong University. He has ever studied in School of Energy and Power Engineering in Xi'an Jiaotong University with the successive degrees of Bachelor, Master, and PhD respectively in energy and power engineering, power engineering and engineering themophysics. He has more than 100 publications in thermal storage, porous media, and heat /mass transfer. He also has good record in educating undergraduates and postgraduates with some approved achievements. He is now serving as the associate editor for special topics and reviews in porous media in Begell House.

Preface to "Recent Advances in Mathematical Aspect in Engineering"

This book contains ten chapters. A total of 53 papers have been submitted for possible publication. After a comprehensive peer review, only 10 papers qualified for final publication. Extensive uses of realistic applications are given in each chapter. For the best understanding of readers, a relevant list of references is also given at the end of each chapter for further study. We wish to thank all reviewers for their excellent suggestions and critical reviews on submitted manuscripts. We were fortunate enough to have prominent scholars who contributed with their original research. We applaud all of them on the successful completion of the Special Issue "Recent Advances in Mathematical Aspects of Engineering". Errors and omissions, if any, are requested to be pointed out, which will be gratefully acknowledged in the next possible Edition. In particular, suggestions for improvement and for the scope and format of the book will be highly appreciated. We express our special gratitude to MPDI for publishing this book. We also want to express our gratitude to the entire editorial team, the contact editor Mrs. Celina Si, and our families and friends for their helpful cooperation. It is worth mentioning that this Special Issue has been cited more than 170 times and viewed more than 11,000 times in a very short span of time. We hope that this book will not only provide an overall picture and the most up-to-date findings from the scientific community working in the field, but also benefit the industrial sectors in specific market niches and end users.

Rahmat Ellahi, Sadiq M. Sait, and Huijin Xu
Editors

Editorial

Recent Advances in Mathematical Aspects of Engineering

Rahmat Ellahi [1,2,*], Sadiq M. Sait [3] and Huijin Xu [4]

1 Department of Mathematics & Statistics, Faculty of Basic and Applied Sciences, International Islamic University, Islamabad 44000, Pakistan
2 Fulbright Fellow Department of Mechanical Engineering, University of California Riverside, Riverside, CA 92521, USA
3 Center for Communications and IT Research, Research Institute, King Fahd University of Petroleum & Minerals, Dhahran 31261, Saudi Arabia; sadiq@kfupm.edu.sa
4 China-UK Low Carbon College, Shanghai Jiao Tong University, Shanghai 200240, China; xuhuijin@sjtu.edu.cn
* Correspondence: rellahi@alumni.ucr.edu

Abstract: This special issue took this opportunity to invite researchers to contribute their latest original research findings, review articles, and short communications on advances in the state of the art of mathematical methods, theoretical studies, or experimental studies that extend the bounds of existing methodologies to new contributions addressing current challenges and engineering problems on "Recent Advances in Mathematical Aspects of Engineering" to be published in *Symmetry*.

Keywords: fluid mechanics; optimization; energy; heat transfer; steady and unsteady flow problems; porosity; nanofluids; particle shape effects; multiphase flow; thermodynamics; magnetohydrodynamics; electromagnetic; physiological fluid phenomena in biological systems; peristaltic; blood flow

Citation: Ellahi, R.; Sait, S.M.; Xu, H. Recent Advances in Mathematical Aspects of Engineering. *Symmetry* **2021**, *13*, 811. https://doi.org/10.3390/sym13050811

Received: 26 January 2021
Accepted: 30 March 2021
Published: 6 May 2021

1. Introduction

In response to a call for papers, a total of 25 papers were submitted for possible publication. After a comprehensive peer review, only 9 papers qualified for acceptance for final publication; the rest 16 papers could not be accommodated. The submissions may have been technically correct but were not considered appropriate for the scope of this special issue. The authors are from 12 geographically distributed countries: the U.S., Mexico, China, Jordan, Saudi Arabia, Pakistan, Malaysia, Vietnam, Taiwan, Thailand, Egypt, and India. This reflects the great impact of the proposed topic and the effective organization of the guest editorial team of this Special Issue. Several theoretical and experimental attempts have been devoted, and this Special Issue is one of them. We hope that this issue will not only address the current challenges but also provide an overall picture and up-to-date findings to readers of the scientific community that ultimately benefits the industrial sector regarding its specific market niches and end users.

2. Methodologies and Usages

The peristaltic flow of a Johnson–Segalman fluid in a symmetric curved channel with convective conditions and flexible walls is addressed in [1]. The channel walls are considered compliant. The main objective of this article is to discuss the effects of a curvilinear channel and heat/mass convection through boundary conditions. The constitutive equations for the Johnson–Segalman fluid are modeled and analyzed under the lubrication approach. The stream function, temperature, and concentration profiles are derived. The analytical solutions are obtained by using the regular perturbation method for a significant number, named the Weissenberg number. The influence of the parameter values on the physical level of interest is outlined and discussed. A comparison is made between Johnson–Segalman and Newtonian fluids. It is concluded that the axial velocity of a Johnson–Segalman fluid is substantially higher than that of a Newtonian fluid.

Bhatti et al. [2] deal with the mass transport phenomena on particle fluid motion through an annulus. A non-Newtonian fluid propagates through a ciliated annulus in the presence of two phenomenon, namely (i) endoscopy and (ii) blood clot. The outer tube is ciliated. To examine the flow behavior, the authors consider the bi-viscosity fluid model. Mathematical modeling is formulated for a small Reynolds number to examine the inertia-free flow. The purpose of this assumption is that the wavelength-to-diameter ratio is maximal, and the pressure can be considerably uniform throughout the entire cross section. The resulting equations are analytically solved, and exact solutions are given for particle- and fluid-phase profiles. The computational software Mathematica is used to evaluate both closed-form and numerical results. The graphical behavior across each parameter is discussed in detail and presented with graphs. The trapping mechanism is also shown across each parameter. It is noted clearly that the particle volume fraction and the blood clot reveal converse behavior on fluid velocity; however, the velocity of the fluid reduces significantly when the fluid behaves as a Newtonian fluid. Schmidt and Soret numbers enhance the concentration mechanism. Furthermore, more pressure is required to pass the fluid when the blood clot appears.

In [3], the problem of finite-time control for nonlinear systems with time-varying delay and exogenous disturbance is studied. First, by constructing a novel augmented Lyapunov–Krasovskii functional involving several symmetric positive definite matrices, a new delay-dependent finite-time boundedness criterion is established for the considered T-S fuzzy time-delay system by employing an improved reciprocally convex combination inequality. Then, a memory state feedback controller is designed to guarantee the finite-time boundness of the closed-loop T-S fuzzy time-delay system, which is in the framework of linear matrix inequalities (LMIs). Finally, the effectiveness and merits of the proposed results are shown by a numerical example.

The maldistribution of fluid flow through multi-channels is a critical issue encountered in many areas, such as multi-channel heat exchangers, electronic device cooling, refrigeration and cryogenic devices, air separation, and the petrochemical industry. The uniformity of flow distribution in a printed circuit heat exchanger (PCHE) is investigated in [4]. The flow distribution and resistance characteristics of a PCHE plate are studied with numerical models under different flow distribution cases. The results show that a sudden change in the angle of the fluid at the inlet of the channel can be greatly reduced by using a spreader plate with an equal inner and outer radius. The flow separation of the fluid at the inlet of the channel can also be weakened, and the imbalance of flow distribution in the channel can be reduced. Therefore, flow uniformity can be improved and the pressure loss between the inlet and outlet of PCHEs can be reduced. The flow maldistribution in each PCHE channel can be reduced to $\pm 0.2\%$, and the average flow maldistribution in all PCHE channels can be reduced to less than 5% when the number of manifolds reaches nine. The numerical simulation of fluid flow distribution can provide guidance for subsequent research and the design and development of multi-channel heat exchangers. In summary, the symmetry of fluid flow in multi-channels for a PCHE is analyzed in this work. This work presents the frequently encountered problem of maldistribution of fluid flow in engineering, and the performance promotion leads to symmetrical aspects in both the structure and the physical process.

The entropy generation on the asymmetric peristaltic propulsion of a non-Newtonian fluid with convective boundary conditions is presented in [5]. The Williamson fluid model is considered for the analysis of flow properties. The current fluid model has the ability to reveal Newtonian and non-Newtonian behavior. The present model is formulated via momentum, entropy, and energy equations, under the approximation of a small Reynolds number and a long wavelength of the peristaltic wave. A regular perturbation scheme is employed to obtain the series solutions up to third-order approximation. All the leading parameters are discussed with the help of graphs for entropy and temperature profiles. The irreversibility process is also discussed with the help of a Bejan number. Streamlines are plotted to examine the trapping phenomena. Results obtained provide an excellent benchmark for further study of the entropy production with mass transfer and a peristaltic pumping mechanism.

In [6], the author examines the unsteady flow over a rotating stretchable disk with deceleration. The highly nonlinear partial differential equations of viscous fluid are simplified by existing similarity transformation. Reduced nonlinear ordinary differential equations are solved by the homotopy analysis method (HAM). The convergence of HAM solutions is also obtained. A comparison table between analytical solutions and numerical solutions is also presented. Finally, the results for useful parameters, i.e., disk stretching parameters and unsteadiness parameters, are found.

The aim of [7] is to examine the rheological significance of a Maxwell fluid configured between two isothermal stretching disks. The energy equation is extended by evaluating the heat source and sink features. The governing partial differential equations (PDEs) are converted to ordinary differential equations (ODEs) by using appropriate variables. An analytically based technique is adopted to compute the series solution of the dimensionless flow problem. The convergence of this series solution is carefully ensured. The physical interpretation of important physical parameters like the Hartmann number, Prandtl number, Archimedes number, Eckert number, heat source/sink parameter, and activation energy parameter are presented for velocity, pressure, and temperature profiles. The numerical values of different involved parameters for the skin friction coefficient and the local Nusselt number are expressed in tabular and graphical form. Moreover, the significance of an important parameter, namely Frank–Kamenetskii, is presented in both tabular and graphical form. This particular study reveals that both axial and radial velocity components decrease by increasing the Frank–Kamenetskii number and stretching the ratio parameter. The pressure distribution is enhanced with an increasing Frank–Kamenetskii number and a stretching ratio parameter. It is also observed that the temperature distribution increases with an increasing Hartmann number, Eckert number, and Archimedes number.

The key objective of the study reported in [8] is to probe the impacts of Brownian motion and thermophoresis diffusion on Casson nanofluid boundary layer flow over a nonlinear inclined stretching sheet, with the effect of convective boundaries and thermal radiations. Nonlinear ordinary differential equations are obtained from governing nonlinear partial differential equations by using compatible similarity transformations. The quantities associated with engineering aspects, such as skin friction, Sherwood number, and heat exchange, along with various impacts of material factors on the momentum, temperature, and concentration, are elucidated and clarified with diagrams. The numerical solution of the present study is obtained via the Keller-box technique and in limiting sense is reduced to the published results for accuracy purpose.

The effects of magnetohydrodynamic 3D nanofluid flow due to a rotating disk subject to Arrhenius activation energy and heat generation/absorption is examined in [9]. Flow is created due to a rotating disk. Velocity, temperature, and concentration slips at the surface of the rotating disk are considered. Effects of thermophoresis and Brownian motion are also accounted. The nonlinear expressions are deduced by the transformation procedure. The shooting technique is used to construct the numerical solution of the governing system. Plots are organized just to investigate how yjr velocity, temperature, and concentration are influenced by various emerging flow parameters. Skin-friction local Nusselt and Sherwood numbers are also plotted and analyzed. In addition, a symmetry is noticed for both components of velocity when the Hartman number increases.

3. Future Trends in Fluid Mechanics

The material that advances the state-of-the-art experimental, numerical, and theoretical methodologies or extends the bounds of existing methodologies through new contributions in symmetry is still insufficient, even with the completion of this Special Issue. The rheological characteristics with thin films under the influence of different nanoparticles and shapes can help with the development of better applications in industry.

Author Contributions: Conceptualization, R.E.; S.M.S.; writing—original draft preparation; formal analysis, H.X. All authors have read and agreed to the published version of the manuscript.

Funding: This research received no external funding.

Institutional Review Board Statement: Not applicable.

Informed Consent Statement: Not applicable.

Data Availability Statement: Not applicable.

Acknowledgments: The guest editorial team of *Symmetry* would like to thank all authors for contributing their original work to this special issue, no matter what the final decision on their submitted manuscript was. The editorial team would also like to thank all anonymous professional reviewers for their valuable time, comments, and suggestions during the review process. We also acknowledge the entire staff of the journal's editorial board for providing their cooperation regarding this Special Issue. We hope that this issue will not only provide an overall picture and most up-to-date findings to readers from the scientific community working in the field but also benefit the industrial sectors in specific market niches and end users.

Conflicts of Interest: The author declares no conflict of interest.

References

1. Yasmin, H.; Iqbal, N.; Hussain, A. Convective Heat/Mass Transfer Analysis on Johnson-Segalman Fluid in a Symmetric Curved Channel with Peristalsis: Engineering Applications. *Symmetry* **2020**, *12*, 1475. [CrossRef]
2. Bhatti, M.M.; Elelamy, A.F.; Sait, S.M.; Ellahi, R. Hydrodynamics Interactions of Metachronal Waves on Particulate-Liquid Motion through a Ciliated Annulus: Application of Bio-Engineering in Blood Clotting and Endoscopy. *Symmetry* **2020**, *12*, 532. [CrossRef]
3. Ruan, Y.; Huang, T. Finite-Time Control for Nonlinear Systems with Time-Varying Delay and Exogenous Disturbance. *Symmetry* **2020**, *12*, 447. [CrossRef]
4. Ke, H.; Lin, Y.; Ke, Z.; Xiao, Q.; Wei, Z.; Chen, K.; Xu, H. Analysis Exploring the Uniformity of Flow Distribution in Multi-Channels for the Application of Printed Circuit Heat Exchangers. *Symmetry* **2020**, *12*, 314. [CrossRef]
5. Riaz, A.; Bhatti, M.M.; Ellahi, R.; Zeeshan, A.; Sait, S.M. Mathematical Analysis on an Asymmetrical Wavy Motion of Blood under the Influence Entropy Generation with Convective Boundary Conditions. *Symmetry* **2020**, *12*, 102. [CrossRef]
6. Sadiq, M.A. Serious Solutions for Unsteady Axisymmetric Flow over a Rotating Stretchable Disk with Deceleration. *Symmetry* **2020**, *12*, 96. [CrossRef]
7. Khan, N.; Nabwey, H.A.; Hashmi, M.S.; Khan, S.U.; Tlili, I. A Theoretical Analysis for Mixed Convection Flow of Maxwell Fluid between Two Infinite Isothermal Stretching Disks with Heat Source/Sink. *Symmetry* **2019**, *12*, 62. [CrossRef]
8. Rafique, K.; Anwar, M.I.; Misiran, M.; Khan, I.; Alharbi, S.O.; Thounthong, P.; Nisar, K.S. Keller-Box Analysis of Buongiorno Model with Brownian and Thermophoretic Diffusion for Casson Nanofluid over an Inclined Surface. *Symmetry* **2019**, *11*, 1370. [CrossRef]
9. Asma, M.; Othman, W.A.M.; Muhammad, T.; Mallawi, F.; Wong, B. Numerical Study for Magnetohydrodynamic Flow of Nanofluid Due to a Rotating Disk with Binary Chemical Reaction and Arrhenius Activation Energy. *Symmetry* **2019**, *11*, 1282. [CrossRef]

Article

Convective Heat/Mass Transfer Analysis on Johnson-Segalman Fluid in a Symmetric Curved Channel with Peristalsis: Engineering Applications

Humaira Yasmin [1,*], Naveed Iqbal [2,*] and Aiesha Hussain [3]

1 Department of Basic Sciences, Preparatory Year Deanship, King Faisal University, Al-Ahsa 31982, Saudi Arabia

2 Department of Mathematics, Faculty of Science, University of Ha'il, Ha'il 81481, Saudi Arabia

3 Department of Mathematics, Quaid-I-Azam University 45320, Islamabad 44000, Pakistan; aieshahussain14@gmail.com

* Correspondence: hhassain@kfu.edu.sa (H.Y.); naveediqbal1989@yahoo.com or n.iqbal@uoh.edu.sa (N.I.)

Received: 19 August 2020; Accepted: 4 September 2020; Published: 8 September 2020

Abstract: The peristaltic flow of Johnson–Segalman fluid in a symmetric curved channel with convective conditions and flexible walls is addressed in this article. The channel walls are considered to be compliant. The main objective of this article is to discuss the effects of curvilinear of the channel and heat/mass convection through boundary conditions. The constitutive equations for Johnson–Segalman fluid are modeled and analyzed under lubrication approach. The stream function, temperature, and concentration profiles are derived. The analytical solutions are obtained by using regular perturbation method for significant number, named as Weissenberg number. The influence of the parameter values on the physical level of interest is outlined and discussed. Comparison is made between Jhonson-Segalman and Newtonian fluid. It is concluded that the axial velocity of Jhonson-Segalman fluid is substantially higher than that of Newtonian fluid.

Keywords: symmetric curved channel; Johnson-Segalman fluid; convective conditions; compliant walls

---

1. Introduction

The researchers have great interest in peristaltic transport of fluids due to immense applications in physiology, biomedical engineering andnindustry. Such motion is caused by a wave of expansion and contraction that propagates along the channel walls. Peristalsis includes the passage of urine from kidney to bladder, swallowing of food through oesophagus, the movement of chyme in the gastrointestinal tract, the vasomotion of small blood vessels, and many others. Blood pumps in the dialysis and heart lung machine operate on the principle of peristaltic action. The roller and finger pumps also operate according to this mechanism. In the nuclear industry, toxic materials can be moved through such a system in order to avoid contaminants from the outside area. Pioneering researches on the topic are presented by Latham [1], Shapiro et al. [2], and Yin and Fung [3]. Currently, abundant literature exists on peristaltic flows of viscous and non-Newtonian fluids under different aspects (see [4–19] and several studies there in). Amongst the several models of non-Newtonian material there is one fluid model that can describe the "spurt" phenomenon. It is subclass of integral type non-Newtonian material and is known as the Johnson–Segalman (JS) fluid. The phrase "spurt" is being used to characterize a significant volume rise to a slight rise in the moving pressure gradient. The contributions of Hayat et al. [20–22] are fundamental in this direction. Elshahed and Haroun [23] investigated the peristaltically moving Johnson–Segalman fluid together with the impact of the magnetism. Wang et al. [24] explored the peristalsis of the Johnson–Segalman fluid across a non-rigid tube. In reality, the configuration of the most physiological tubes and glandular ducts is curved.

In this context, the effect of curvature appears to be meaningful. This fact gives great motivation to study peristaltic flow through curved channels. In the first place, Sato et al. [25] addressed the two-dimensional peristaltic transport of viscous liquid inside a curved channel. Ali et al. [26] revisited the analysis of Sato et al. [25] in a wave frame. Some more interesting studies for peristalsis in a curved channel can be consulted through [27–31].

The effect of heat transfer has vast applications in food processing, dilation of blood vessels, heat conduction in tissues, and its convection due to blood flow from the pores of the tissues. The impact of both heat and mass transfer plays an essential part in spreading of chemical pollutants in saturated soil, underground disposal of nuclear waste, thermal insulation, enhanced oil recovery, etc. The effects of mass transfer arose in diffusion, combustion, and distillation processes, and in many other industrial processes. Convective heat transfer through boundary conditions is used in systems, such as steam turbines, nuclear power stations, thermal energy storage, etc. In this context, Hina and Hayat [32] examined the effects of heat/mass transfer on Johnson-Segalman liquid inducing peristaltic movement in a compliant curved channel. Mehmood et al. [33], Hayat et al. [34] and Riaz et al. [35] analyzed the characteristics of heat flux in peristaltic transport with/without compliant walls. Hayat et al. [36–39] conducted an analysis of non-Newtonian fluids with peristalsis in the presence of convective constraints. Yasmin et al. [40] discussed the effects of convective conditions in peristalsis of Johnson–Segalman fluid in an asymmetric channel.

It is noted that the peristalsis of non-Newtonian fluid in a curved channel with convective mass transfer conditions at the walls is not addressed so far. Even such analysis is not carried out for viscous fluids. The current research paper varies from the existing results in terms of convective boundary conditions. The key focus of this paper is the implementation of a novel definition of convective heat and mass transfer conditions in the theory of Johnson–Segalman fluid transferred via a peristaltic motion across a curved channel. Hence, in this attempt, the convective conditions for both heat and mass transfer are considered. An incompressible Johnson–Segalman fluid is considered in a curved channel. The set of solutions for the small value of Weissenberg number are developed. The obtained results are visualized and thoroughly analyzed. Impacts reflecting the influence of pertinent parameters are pointed out physically.

2. Problem Formulation

We anticipate the peristaltic transport of the incompressible Johnson–Segalman fluid in a symmetric curved half-width (d_1) channel clasped in a circular pattern with center O and radius R^* (see Figure 1 and Ref. [32]).

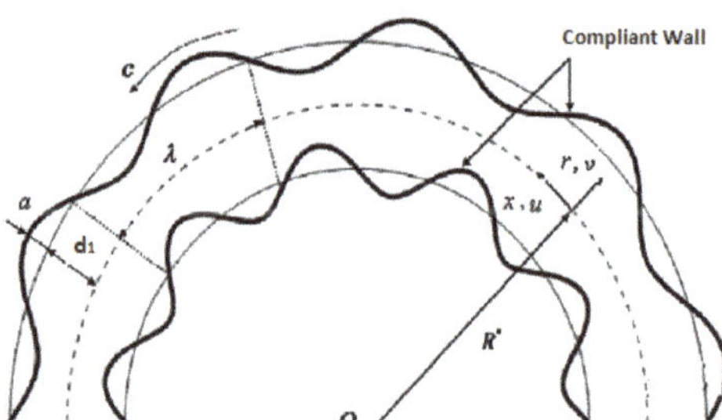

Figure 1. Schematic diagram of the problem.

The flow in the channel is stimulated by small amplitude sinusoidal waves that travel along the compliant walls. The axial direction of the flow is x and r is radial direction. Here, v and u are the velocity vector components in the radial and axial directions, respectively. The wave shapes at channel walls are considered symmetric and given by

$$r = \pm\eta(x,t) = \pm\left[d_1 + a\,\sin\left(\frac{2\,\pi}{\lambda}\,(x - c\,t)\right)\right], \tag{1}$$

where c is the wave speed and λ is the wavelength, respectively.

The continuity and momentum equations governing the flow can be written as [32]:

$$\frac{\partial[(r+R^*)v]}{\partial r} + R^*\frac{\partial u}{\partial x} = 0, \tag{2}$$

$$\rho\left(\frac{\partial v}{\partial t} + v\frac{\partial v}{\partial r} + \frac{R^*u}{r+R^*}\frac{\partial v}{\partial x} - \frac{u^2}{r+R^*}\right) = -\frac{\partial p}{\partial r} + \frac{1}{r+R^*}\frac{\partial}{\partial r}[(r+R^*)\tau_{rr}] + \frac{R^*}{r+R^*}\frac{\partial \tau_{xr}}{\partial x} - \frac{\tau_{xx}}{r+R^*}, \tag{3}$$

$$\rho\left(\frac{\partial u}{\partial t} + v\frac{\partial u}{\partial r} + \frac{R^*u}{r+R^*}\frac{\partial u}{\partial x} + \frac{uv}{r+R^*}\right) = -\frac{R^*}{r+R^*}\frac{\partial p}{\partial x} + \frac{1}{(r+R^*)^2}\frac{\partial}{\partial r}[(r+R^*)^2\tau_{rx}] + \frac{R^*}{r+R^*}\frac{\partial \tau_{xx}}{\partial x}. \tag{4}$$

The equations for energy and concentration [32] are given by

$$\begin{aligned}\rho C_p\left(\frac{\partial T}{\partial t} + v\frac{\partial T}{\partial r} + \frac{R^*u}{r+R^*}\frac{\partial T}{\partial x}\right) &= \kappa\left(\frac{\partial^2 T}{\partial x^2}\left(\frac{R^*}{r+R^*}\right)^2 + \frac{1}{r+R^*}\frac{\partial T}{\partial r} + \frac{\partial^2 T}{\partial r^2}\right) + (S_{rr} - S_{xx})\frac{\partial v}{\partial r}\\ &\quad + S_{xr}\left(\frac{\partial u}{\partial r} + \frac{R^*}{r+R^*}\frac{\partial v}{\partial x} - \frac{u}{r+R^*}\right),\end{aligned} \tag{5}$$

$$\begin{aligned}\frac{\partial C}{\partial t} + v\frac{\partial C}{\partial r} + \frac{R^*u}{r+R^*}\frac{\partial C}{\partial x} &= D\left(\frac{\partial^2 C}{\partial x^2}\left(\frac{R^*}{r+R^*}\right)^2 + \frac{1}{r+R^*}\frac{\partial C}{\partial r} + \frac{\partial^2 C}{\partial r^2}\right)\\ &\quad + \frac{DK_T}{T_m}\left(\frac{\partial^2 T}{\partial r^2} + \frac{1}{r+R^*}\frac{\partial T}{\partial r} + \left(\frac{R^*}{r+R^*}\right)^2\frac{\partial^2 T}{\partial x^2}\right).\end{aligned} \tag{6}$$

For the Johnson-Segalman fluid, the stress tensor $\mathbf{ø}$ is

$$\mathbf{ø} = 2\mu\mathbf{D} + \mathbf{S},$$

in which the extra stress tensor $\mathbf{S}$ needs to satisfy the relationship

$$\mathbf{S} + m\left(\frac{d\mathbf{S}}{dt} + \mathbf{S}(\mathbf{W} - \xi\mathbf{D}) + (\mathbf{W} - \xi\mathbf{D})^T\mathbf{S}\right) = 2\eta_1\mathbf{D},$$

where

$$\mathbf{D} = \frac{[(grad\mathbf{V})^T + grad\mathbf{V}]}{2},$$

$$\mathbf{W} = \frac{[grad\mathbf{V} - (grad\mathbf{V})^T]}{2}.$$

The relations listed above produce the following equations:

$$S_{rr} + m\left[\frac{dS_{rr}}{dt} - \frac{2uS_{rx}}{r+R^*} + S_{rx}\left\{(1-\xi)\frac{\partial u}{\partial r} - \frac{1+\xi}{r+R^*}[R^*\frac{\partial v}{\partial x} - u]\right\} - 2\xi S_{rr}\frac{\partial v}{\partial r}\right] = 2\eta_1\frac{\partial v}{\partial r}, \tag{7}$$

$$\begin{aligned}&S_{rx} + m\frac{dS_{rx}}{dt} + \frac{mu(S_{rr} - S_{xx})}{r+R^*} + \frac{mS_{xx}}{2}\left\{(1-\xi)\frac{\partial u}{\partial r} - \frac{1+\xi}{r+R^*}\left[R^*\frac{\partial v}{\partial x} - u\right]\right\}\\ &+\frac{mS_{rr}}{2}\left\{\frac{1-\xi}{r+R^*}\left[R^*\frac{\partial v}{\partial x} - u\right] - (1+\xi)\frac{\partial u}{\partial r}\right\}\\ =\;& \eta_1\left(\frac{\partial u}{\partial r} + \frac{R^*}{r+R^*}\frac{\partial v}{\partial x} - \frac{u}{r+R^*}\right),\end{aligned} \tag{8}$$

$$S_{xx}+m\left[\frac{dS_{xx}}{dt}+\frac{2uS_{rx}}{r+R^*}-S_{rx}\left\{(1+\xi)\frac{\partial u}{\partial r}-\frac{1-\xi}{r+R^*}\left[R^*\frac{\partial v}{\partial x}-u\right]\right\}+2\xi S_{xx}\frac{\partial v}{\partial r}\right]=-2\eta_1\frac{\partial v}{\partial r}, \quad (9)$$

where $\frac{d}{dt}=\frac{\partial}{\partial t}+v\frac{\partial}{\partial r}+\frac{R^*u}{r+R^*}\frac{\partial}{\partial x}$ represents the material derivative with respect to time, ρ denotes the fluid density, κ the thermal conductivity of fluid, **W** and **D** skew-symmetric and symmetrical parts of the gradient of velocity, ξ the slip parameter, C_p the fluid specific heat, T and C are the fluid temperature and concentration, respectively, the thermal diffusion ratio is K_T, D the mass diffusivity coefficient, T_m represents the mean/average temperature, μ and η_1 the viscosities, and m the relaxation time.

The appropriate boundary conditions are

$$u=0 \ \text{ at } \ r=\pm\eta, \quad (10)$$

$$k\frac{\partial T}{\partial r} = -h_1(T-T_0) \ \text{ at } \ r=+\eta, \quad (11)$$

$$k\frac{\partial T}{\partial r} = -h_2(T_0-T) \ \text{ at } \ r=-\eta, \quad (12)$$

$$D\frac{\partial C}{\partial r} = -h_3(C-C_0) \ \text{ at } \ r=+\eta, \quad (13)$$

$$D\frac{\partial C}{\partial r} = -h_4(C_0-C) \ \text{ at } \ r=-\eta, \quad (14)$$

$$\begin{aligned} R^*[-\tau\frac{\partial^3}{\partial x^3}+m_1\frac{\partial^3}{\partial x\partial t^2}+d\frac{\partial^2}{\partial t\partial x}]\eta &= \frac{1}{r+R^*}\frac{\partial}{\partial r}\{(r+R^*)^2\tau_{rx}\}+R^*\frac{\partial\tau_{xx}}{\partial x}-\rho(r+R^*) \\ \left[\frac{\partial u}{\partial t}+v\frac{\partial u}{\partial r}+\frac{R^*u}{r+R^*}\frac{\partial u}{\partial x}+\frac{uv}{r+R^*}\right] \text{ at } r &= \pm\eta. \end{aligned} \quad (15)$$

Here, the pressure, the time, the fluid density, and the curvature parameters are p, t, ρ, and R^*, respectively, T_0 and C_0 the ambient temperature and concentration, h_1 and h_2 the coefficients of heat transfer at upper and lower walls, h_3 and h_4 the coefficients of mass transfer at upper and lower walls, S_{rr}, S_{rx}, S_{xr} and S_{xx} the components of the extra stress tensor **S**, τ the elastic tension, d the viscous damping coefficient, and m_1 the mass per unit area. Equation (10) is the no slip condition for velocity profile. Equations (11) and (12) are the convective boundary conditions for heat transfer. Analogues to the convective heat transfer at the boundary, we also use the mixed condition for the mass transfer as well (i.e., Equations (13) and (14)).

Employing the aforementioned dimensionless variables

$$\begin{aligned} x^* &= \frac{x}{\lambda}, r^*=\frac{r}{d_1}, u^*=\frac{u}{c}, v^*=\frac{v}{c}, \Psi^*=\frac{\Psi}{cd_1}, t^*=\frac{ct}{\lambda}, \\ \eta^* &= \frac{\eta}{d_1}, k=\frac{R^*}{d_1}, p^*=\frac{d_1^2 p}{c\lambda(\mu+\eta_1)}, \varepsilon=\frac{a}{d}, \delta=\frac{d_1}{\lambda}, \\ \theta &= \frac{T-T_0}{T_0}, \phi=\frac{C-C_0}{C_0}, S_{ij}^*=\frac{d_1 S_{ij}}{c\eta_1}, We=\frac{mc}{d_1}, \end{aligned}$$

Equations (7)–(9) become

$$\begin{aligned} 2\frac{\partial v}{\partial r} &= S_{rr}+We\left[\left(\delta\frac{\partial}{\partial t}+v\frac{\partial}{\partial r}+\frac{uk\delta}{r+k}\frac{\partial}{\partial x}\right)S_{rr}-\frac{2uS_{rx}}{r+k}-2\xi S_{rr}\frac{\partial v}{\partial r}\right] \\ &+WeS_{rx}\left\{(1-\xi)\frac{\partial u}{\partial r}-\frac{1+\xi}{r+k}\left(k\delta\frac{\partial v}{\partial x}-u\right)\right\}, \end{aligned} \quad (16)$$

$$\begin{aligned}\left(\frac{\partial u}{\partial r}+\frac{k\delta}{r+k}\frac{\partial v}{\partial x}-\frac{u}{r+k}\right) &= S_{rx}+We\left[\left(\delta\frac{\partial}{\partial t}+v\frac{\partial}{\partial r}+\frac{uk\delta}{r+k}\frac{\partial}{\partial x}\right)S_{rx}+\frac{u(S_{rr}-S_{xx})}{r+k}\right]\\ &+\frac{WeS_{rr}}{2}\left\{\frac{1-\xi}{r+k}\left[k\delta\frac{\partial v}{\partial x}-u\right]\right\}-(1+\xi)\frac{\partial u}{\partial r}\\ &+\frac{WeS_{xx}}{2}\left\{(1-\xi)\frac{\partial u}{\partial r}-\frac{1+\xi}{r+k}\left[k\delta\frac{\partial v}{\partial x}-u\right]\right\},\end{aligned} \tag{17}$$

$$\begin{aligned}-2\frac{\partial v}{\partial r} &= S_{xx}+We\left[\left(\delta\frac{\partial}{\partial t}+v\frac{\partial}{\partial r}+\frac{uk\delta}{r+k}\frac{\partial}{\partial x}\right)S_{xx}+\frac{2uS_{rx}}{r+k}-2\xi S_{xx}\frac{\partial v}{\partial r}\right]\\ &+WeS_{rx}\left\{\frac{1-\xi}{r+k}\left(k\delta\frac{\partial v}{\partial x}-u\right)-(1+\xi)\frac{\partial u}{\partial r}\right\},\end{aligned} \tag{18}$$

and Equations (4)–(6) are reduced to

$$\begin{aligned}\mathrm{Re}\,\delta\left[\delta\frac{\partial v}{\partial t}+v\frac{\partial v}{\partial r}+\frac{\partial v}{\partial x}\frac{k\delta u}{r+k}-\frac{u^2}{r+k}\right] &= -\frac{\eta_1+\mu}{\eta_1}\frac{\partial p}{\partial r}+\frac{4\delta\mu}{\eta_1(r+k)}\frac{\partial v}{\partial r}+\frac{k\delta^3}{r+k}\frac{\partial S_{rx}}{\partial x}+\delta\frac{\partial S_{rr}}{\partial r}\\ &+\frac{\delta(S_{rr}-S_{xx})}{r+k}+\frac{\delta\mu}{\eta_1}\frac{\partial^2 v}{\partial r^2}+\frac{\delta^2 k\mu}{\eta_1(r+k)}\\ &\times\frac{\partial}{\partial x}\left(\frac{\partial u}{\partial r}+\frac{k\delta}{r+k}\frac{\partial v}{\partial x}-\frac{u}{r+k}\right),\end{aligned} \tag{19}$$

$$\begin{aligned}\mathrm{Re}\left[\delta\frac{\partial u}{\partial t}+v\frac{\partial u}{\partial r}+\frac{k\delta u}{r+k}\frac{\partial u}{\partial x}-\frac{uv}{r+k}\right] &= -\frac{\eta_1+\mu}{\eta_1(r+k)}\frac{\partial p}{\partial x}+\frac{2S_{rx}}{r+k}+\frac{\partial S_{rx}}{\partial r}+\frac{k\delta}{r+k}\frac{\partial S_{xx}}{\partial x}\\ &-\frac{2k\delta\mu}{(r+k)\eta_1}\times\frac{\partial^2 v}{\partial r\partial x}+\frac{\mu}{\eta_1}\frac{\partial}{\partial x}\left(\frac{\partial u}{\partial r}+\frac{k\delta}{r+k}\frac{\partial v}{\partial x}-\frac{u}{r+k}\right)\\ &+\frac{\delta\mu}{\eta_1}\frac{\partial^2 v}{\partial r^2}+\frac{\delta^2 k\mu}{\eta_1(r+k)}\times\frac{\partial}{\partial r}\left(\frac{\partial u}{\partial r}+\frac{k\delta}{r+k}\frac{\partial v}{\partial x}-\frac{u}{r+k}\right)\\ &+\frac{2\mu}{\eta_1(r+k)}\left(\frac{\partial u}{\partial r}+\frac{\partial v}{\partial x}\frac{k\delta}{r+k}-\frac{u}{r+k}\right),\end{aligned} \tag{20}$$

$$\begin{aligned}\mathrm{Re}\left[\delta\frac{\partial\theta}{\partial t}+v\frac{\partial\theta}{\partial r}+\frac{\partial\theta}{\partial x}\frac{k\delta u}{r+k}\right] &= E\left[S_{xr}\left(\frac{\partial u}{\partial r}+\frac{k\delta}{r+k}\frac{\partial v}{\partial x}-\frac{u}{r+k}\right)+(S_{rr}-S_{xx})\frac{\partial v}{\partial r}\right]\\ &+\frac{1}{\mathrm{Pr}}\left[\frac{\partial^2\theta}{\partial r^2}+\frac{1}{r+k}\frac{\partial\theta}{\partial r}+\delta^2\frac{\partial^2\theta}{\partial x^2}\right],\end{aligned} \tag{21}$$

$$\begin{aligned}\mathrm{Re}\left[\delta\frac{\partial\phi}{\partial t}+v\frac{\partial\phi}{\partial r}+\frac{k\delta u}{r+k}\frac{\partial\phi}{\partial x}\right] &= \frac{1}{Sc}\left[\frac{\partial^2\phi}{\partial r^2}+\frac{1}{r+k}\frac{\partial\phi}{\partial r}+\delta^2\frac{\partial^2\phi}{\partial x^2}\right]\\ &+Sr\left[\frac{\partial^2\theta}{\partial r^2}+\frac{1}{r+k}\frac{\partial\theta}{\partial r}+\delta^2\frac{\partial^2\theta}{\partial x^2}\right],\end{aligned} \tag{22}$$

with

$$u=0 \quad \text{at} \quad r=\pm\eta, \tag{23}$$

$$\frac{\partial\theta}{\partial r}+Bi_1\theta = 0 \quad \text{at} \quad r=+\eta, \tag{24}$$

$$\frac{\partial\theta}{\partial r}-Bi_2\theta = 0 \quad \text{at} \quad r=-\eta, \tag{25}$$

$$\frac{\partial\varphi}{\partial r}+Bi_3\phi = 0 \quad \text{at} \quad r=+\eta, \tag{26}$$

$$\frac{\partial\varphi}{\partial r}-Bi_4\phi = 0 \quad \text{at} \quad r=-\eta, \tag{27}$$

$$
\begin{aligned}
k[E_1\frac{\partial^3}{\partial x^3}+E_2\frac{\partial^3}{\partial x\partial t^2}+E_3\frac{\partial^2}{\partial t\partial x}]\eta \quad = \quad & \frac{\eta_1(r+k)}{\eta_1+\mu}\left[\frac{\partial}{\partial r}\left(\frac{\partial u}{\partial r}+\frac{k\delta}{r+k}\frac{\partial v}{\partial x}-\frac{u}{r+k}\right)-\frac{2k\delta}{(r+k)}\frac{\partial^2 v}{\partial r\partial x}\right] \\
& -\frac{R_e\mu(r+k)}{\eta_{1+}\mu}\left[\delta\frac{\partial u}{\partial t}+v\frac{\partial u}{\partial r}+\frac{k\delta u}{r+k}\frac{\partial u}{\partial x}+\frac{uv}{r+k}\right] \\
& +\frac{\eta_1(r+k)}{\eta_1+\mu}\left[\frac{\partial S_{rx}}{\partial r}+\frac{\partial S_{rx}}{\partial x}\frac{k\delta}{r+k}+\frac{2S_{rx}}{r+k}\right]+ \\
& \frac{2\mu}{(\eta_1+\mu)}\left(\frac{\partial u}{\partial r}+\frac{\partial v}{\partial x}\frac{k\delta}{r+k}-\frac{u}{r+k}\right) \quad \text{at } r\pm\eta.
\end{aligned} \tag{28}
$$

Defining the stream function $\psi(x,r,t)$ by

$$
u=-\frac{\partial\psi}{\partial r}, v=\delta\frac{k}{r+k}\frac{\partial\psi}{\partial x}, \tag{29}
$$

Equation (2) is automatically satisfied and Equations (16)–(28) subject to lubrication approach become

$$
0=S_{rr}+WeS_{rx}\left[-(1-\xi)\,\psi_{rr}-\frac{1+\xi}{r+k}\psi_r+\frac{2\psi_r}{r+k}\right], \tag{30}
$$

$$
\begin{aligned}
\left(-\psi_{rr}+\frac{\psi_r}{r+k}\right) \quad = \quad & S_{rx}-We\frac{\psi_r(S_{rr}-S_{xx})}{r+k} \\
& +\frac{WeS_{rr}}{2}\left\{\frac{1-\xi}{r+k}\psi_r+(1+\xi)\,\psi_{rr}\right\} \\
& -\frac{WeS_{xx}}{2}\left\{\frac{1+\xi}{r+k}\psi_r+(1-\xi)\,\psi_{rr}\right\},
\end{aligned} \tag{31}
$$

$$
0=S_{xx}+WeS_{rx}\left[(1+\xi)\,\psi_{rr}+\frac{1-\xi}{r+k}\psi_r-\frac{2\psi_r}{r+k}\right], \tag{32}
$$

$$
\frac{\partial p}{\partial r}=0, \tag{33}
$$

$$
-\frac{k(\eta_1+\mu)}{\eta_1\,(r+k)}\frac{\partial p}{\partial x}+\frac{\partial S_{rx}}{\partial r}+\frac{2S_{rx}}{r+k}+\frac{\mu}{\eta_1}\frac{\partial}{\partial r}\left(-\psi_{rr}+\frac{\psi_r}{r+k}\right)+\frac{2\mu}{\eta_1\,(r+k)}\left(-\psi_{rr}+\frac{\psi_r}{r+k}\right)=0, \tag{34}
$$

$$
\left[\frac{\partial^2}{\partial r^2}+\frac{1}{k+r}\frac{\partial}{\partial r}\right]\theta=-Br\left[S_{rx}\left(-\psi_{rr}+\frac{\psi_r}{k+r}\right)\right], \tag{35}
$$

$$
\left[\frac{\partial^2}{\partial r^2}+\frac{1}{k+r}\frac{\partial}{\partial r}\right]\phi=-ScSr\left[\frac{\partial^2}{\partial r^2}+\frac{1}{k+r}\frac{\partial}{\partial r}\right]\theta, \tag{36}
$$

$$
\psi_r=0\;atr=\pm\eta=\pm\left[1+\varepsilon\sin 2\pi(x-t)\right], \tag{37}
$$

$$
\frac{\partial\theta}{\partial r}+Bi_1\theta=0 \quad \text{at} \quad r=+\eta, \tag{38}
$$

$$
\frac{\partial\theta}{\partial r}-Bi_2\theta=0 \quad \text{at} \quad r=-\eta, \tag{39}
$$

$$
\frac{\partial\varphi}{\partial r}+Bi_3\phi=0 \quad \text{at} \quad r=+\eta, \tag{40}
$$

$$
\frac{\partial\varphi}{\partial r}-Bi_4\phi=0 \quad \text{at} \quad r=-\eta, \tag{41}
$$

$$
\begin{aligned}
k\left[E_1\frac{\partial^3}{\partial x^3}+E_2\frac{\partial^3}{\partial x\partial t^2}+E_3\frac{\partial^2}{\partial x\partial t}\right]\eta \quad = \quad & \frac{\eta_1(r+k)}{\eta_1+\mu}\left[\frac{\mu}{\eta_1}\frac{\partial}{\partial r}\left(-\psi_{rr}+\frac{\psi_r}{k+r}\right)\right]+\frac{\partial S_{rx}}{\partial r}+\frac{2S_{rx}}{r+k} \\
+\frac{2\mu}{\eta_1+\mu}\left(-\psi_{rr}+\frac{\psi_r}{r+k}\right) \text{ at } r \quad = \quad & \pm\eta,
\end{aligned} \tag{42}
$$

where the amplitude ratio is represented by $\epsilon(= a/d_1)$, $\delta(= d_1/\lambda)$ is the wave number, the dimensionless curvature parameter is k, $E_1 = -\frac{\tau d_1^3}{\lambda^3 \eta_1 c}$, $E_2 = \frac{m_1 c d_1^3}{\lambda^3 \eta_1 c}$, $E_3 = \frac{d d_1^3}{\lambda^2 \eta_1}$ the non-dimensional elasticity parameters, $R_e = \frac{c\rho d_1}{\eta_1 \lambda^2}$ the Reynolds number, $We = mc/d_1$ the Weissenberg number, the Prandtl number is denoted by $Pr = \mu C_p/\kappa$, the Eckert number is $E = c^2/CpT_0$, the Schmidt numbers is $Sc = \mu/\rho D$, the Soret number is $Sr(= \rho T_0 D K_T/\mu T_m C_0)$, $EPr = Br$ is the Brinkman number, and $Bi_1 = h_1 d/k$, $Bi_2 = h_2 d/k$, $Bi_3 = h_3 d/D$ and $Bi_4 = h_4 d/D$ the Biot numbers for heat/mass transfer.

From Equations (30)–(32), one can get

$$S_{rx} = \left(-\psi_{rr} + \frac{\psi_r}{r+k}\right)\left[1 + We^2\left(1-\xi^2\right)\left(-\psi_{rr} + \frac{\psi_r}{r+k}\right)^2\right]^{-1}. \tag{43}$$

Additionally, Equations (33) and (34) give

$$(r+k)\frac{\partial^2 S_{rx}}{\partial r^2} + 3\frac{\partial S_{rx}}{\partial r} + \frac{(k+r)\mu}{\eta_1}\frac{\partial^2}{\partial r^2}\left(-\psi_{rr} + \frac{\psi_r}{r+k}\right) + \frac{3\mu}{\eta_1}\frac{\partial}{\partial r}\left(-\psi_{rr} + \frac{\psi_r}{k+r}\right) = 0. \tag{44}$$

Heat transfer coefficient at the wall is defined by

$$Z = \eta_x \theta_y(\eta). \tag{45}$$

3. Method of Solution

We have used the standard perturbation approach relying on a small parameter to solve the strictly nonlinear differential equations, because the exact solution is not achievable. This approach is helpful in finding an approximate solution to the problem, beginning with an exact solution to a similar and simplified problem. This approach is more efficient, as it provides a solution in the form of a converging series. In order to find the series solution of the problem, we expand ψ, p and S_{rx} in terms of small parameter We^2. Therefore, we can write the flow quantities, as follows:

$$\psi = \psi_0 + We^2\psi_1 + \ldots, \tag{46}$$
$$S_{rx} = S_{0rx} + We^2 S_{1rx} + \ldots, \tag{47}$$
$$S_{rr} = S_{0rr} + We^2 S_{1rr} + \ldots, \tag{48}$$
$$S_{xx} = S_{0xx} + We^2 S_{1xx} + \ldots, \tag{49}$$
$$\theta = \theta_0 + We^2\theta_1 + \ldots, \tag{50}$$
$$\phi = \phi_0 + We^2\phi_1 + \ldots, \tag{51}$$
$$Z = Z_0 + We^2 Z_1 + \ldots. \tag{52}$$

4. Results

4.1. Zeroth Order System

Using Equations (46)–(52) into Equations (35)–(45) and then equating the coefficients of We^0 we have

$$(k+r)\frac{\partial^2}{\partial r^2}\left(-\psi_{0rr} + \frac{\psi_{0r}}{k+r}\right) + 3\frac{\partial}{\partial r}\left(-\psi_{0rr} + \frac{\psi_{0r}}{k+r}\right) = 0, \tag{53}$$

$$\left[\frac{\partial^2}{\partial r^2} + \frac{1}{k+r}\frac{\partial}{\partial r}\right]\theta_0 = -Br\left[S_{0rx}\left(-\psi_{0rr} + \frac{\psi_{0r}}{k+r}\right)\right], \tag{54}$$

$$\left[\frac{\partial^2}{\partial r^2} + \frac{1}{k+r}\frac{\partial}{\partial r}\right]\phi_0 = -ScSr\left[\frac{\partial^2}{\partial r^2} + \frac{1}{k+r}\frac{\partial}{\partial r}\right]\theta_0 \tag{55}$$

$$\psi_{0r} = 0, \text{ at } r = \pm\eta, \tag{56}$$

$$\frac{\partial \theta_0}{\partial r} + Bi_1\theta_0 = 0 \quad \text{at} \quad r = +\eta, \tag{57}$$

$$\frac{\partial \theta_0}{\partial r} - Bi_2\theta_0 = 0 \quad \text{at} \quad r = -\eta, \tag{58}$$

$$\frac{\partial \varphi_0}{\partial r} + Bi_3\phi_0 = 0 \quad \text{at} \quad r = +\eta, \tag{59}$$

$$\frac{\partial \varphi_0}{\partial r} - Bi_4\phi_0 = 0 \quad \text{at} \quad r = -\eta, \tag{60}$$

$$k\left[E_1\frac{\partial^3\eta}{\partial x^3} + E_2\frac{\partial^3\eta}{\partial x\partial t^2} + E_3\frac{\partial^2\eta}{\partial x\partial t}\right] = (r+k)\frac{\partial}{\partial r}\left(-\psi_{0rr} + \frac{\psi_{0r}}{k+r}\right) + 2\left(-\psi_{0rr} + \frac{\psi_{0r}}{k+r}\right), \text{ at } r = \pm\eta, \tag{61}$$

where

$$S_{0rx} = \left(-\psi_{0rr} + \frac{\psi_{0r}}{r+k}\right).$$

Solving Equations (53)–(55), we get

$$\psi_0 = C_1 + C_2\ln(r+k) + C_3(r+k)^2 + C_4(r+k)^2\ln(r+k), \tag{62}$$

$$\theta_0 = A_1 + A_2\ln(r+k) + 4BrC_2C_4(\ln(r+k))^2 - Br\left(C_4(r+k)^2 + \frac{C_2^2}{(r+k)^2}\right), \tag{63}$$

$$\phi_0 = B_1\ln(r+k) + B_2 + \frac{BrC_2^2ScSr}{(k+r)^2} + BrC_4^2(k+r)^2ScSr - 4BrC_2C_4ScSr(\ln(r+k))^2, \tag{64}$$

and heat transfer coefficient is given by

$$Z_0 = \eta_x\left(\frac{A_2}{k+\eta} + Br\left(\frac{2C_2^2}{(k+\eta)^3} - 2C_4^2(k+\eta)\right) + \frac{8BrC_2C_4\ln(k+\eta)}{k+\eta}\right), \tag{65}$$

where

$$C_1 = 0,$$

$$C_2 = -L(k^2-\eta^2)^2(\ln(k+\eta) - \ln(k-\eta)),$$

$$C_3 = \frac{L(2k\eta + (k+\eta)^2\ln(k+\eta) - (k-\eta)^2\ln(k-\eta))}{16k\eta},$$

$$C_4 = -\frac{L}{4},$$

$$A_1 = \frac{BrM_1(M_{10} + M_5 - M_7 - C_4^2M_9) - BrM(M_4 + M_6 + C_4^2M_8 - M_{11})}{M_3},$$

$$A_2 = \frac{Bi_1BrM_{12} + BrM_{13} + M_{14} - \frac{8BrC_2C_4\ln(k+\eta)}{k+\eta} - 4Bi_1BrC_2C_4\ln(k+\eta)^2}{M},$$

$$B_1 = \frac{BrScSr(2Bi_4Y_1 + 4C_2C_4(Y_2 - Y_4 + Y_5) + 2Bi_3(C_2^2Y_3 + C_4^2(\eta - k(1+2Bi_4\eta))))}{Y},$$

$$B_2 = \frac{BrScSr(M_5 + Y_9 + \frac{2C_2^2}{(k+\eta)^3} - \frac{Bi_3C_2^2}{(k+\eta)^2} - 2C_4^2(k+\eta) - Bi_3C_4^2(k+\eta))}{Bi_3}.$$

4.2. First Order System

The coefficients of $O(We^2)$ form the following expressions:

$$\begin{aligned} 0 &= (r+k)\frac{\partial^2}{\partial r^2}\left[\left(-\psi_{1rr}+\frac{\psi_{1r}}{r+k}\right)-\frac{(1-\xi^2)\,\eta_1}{(\eta_1+\mu)}\left(-\psi_{0rr}+\frac{\psi_{0r}}{r+k}\right)^3\right] \\ &+3\frac{\partial}{\partial r}\left[\left(-\psi_{1rr}+\frac{\psi_{1r}}{r+k}\right)-\frac{(1-\xi^2)\,\eta_1}{(\eta_1+\mu)}\left(-\psi_{0rr}+\frac{\psi_{0r}}{r+k}\right)^3\right], \end{aligned} \tag{66}$$

$$\left[\frac{\partial^2}{\partial r^2}+\frac{1}{k+r}\frac{\partial}{\partial r}\right]\theta_1=-Br\left[S_{1rx}\left(-\psi_{0rr}+\frac{\psi_{0r}}{k+r}\right)+S_{0rx}\left(-\psi_{1rr}+\frac{\psi_{1r}}{k+r}\right)\right], \tag{67}$$

$$\left[\frac{\partial^2}{\partial r^2}+\frac{1}{k+r}\frac{\partial}{\partial r}\right]\phi_1=-ScSr\left[\frac{\partial^2}{\partial r^2}+\frac{1}{k+r}\frac{\partial}{\partial r}\right]\theta_1, \tag{68}$$

$$\psi_{1r}=0,\ \text{at}\ r=\pm\eta, \tag{69}$$

$$\frac{\partial\theta_1}{\partial r}+Bi_1\theta_1=0,\ \text{at}\ r=+\eta, \tag{70}$$

$$\frac{\partial\theta_1}{\partial r}-Bi_2\theta_1=0,\ \text{at}\ r=-\eta, \tag{71}$$

$$\frac{\partial\phi_1}{\partial r}+Bi_3\phi_1=0,\ \text{at}\ r=+\eta, \tag{72}$$

$$\frac{\partial\phi_1}{\partial r}-Bi_4\phi_1=0,\ \text{at}\ r=-\eta, \tag{73}$$

$$\begin{aligned} 0 &= (r+k)\frac{\partial}{\partial r}\left[\left(-\psi_{1rr}+\frac{\psi_{1r}}{r+k}\right)-\frac{(1-\xi^2)\,\eta_1}{(\eta_1+\mu)}\left(-\psi_{0rr}+\frac{\psi_{0r}}{r+k}\right)^3\right] \\ &+2\left[\left(-\psi_{1rr}+\frac{\psi_{1r}}{r+k}\right)-\frac{(1-\xi^2)\,\eta_1}{(\eta_1+\mu)}\left(-\psi_{0rr}+\frac{\psi_{0r}}{r+k}\right)^3\right],\ \text{at}\ r=\pm\eta, \end{aligned} \tag{74}$$

with

$$S_{1rx}=\left(-\psi_{1rr}+\frac{\psi_{1r}}{r+k}\right)-\left(1-\xi^2\right)\left(-\psi_{0rr}+\frac{\psi_{0r}}{r+k}\right)^3. \tag{75}$$

The results corresponding to the first order are

$$\begin{aligned}
\psi_1 &= 1/3(k+r)^4(\eta_1+\mu)[C_2^3\eta_1(-1+\xi^2)] - 1/(k+r)^2(\eta_1+\mu)[3C_2^2C_4\eta_1(-1+\xi^2)] \\
&\quad +krC_{12} + 1/2r^2C_{12} - 1/4(k+r)^2C_{13} \\
&\quad +C_{14} + C_{11}\log(k+r) + 1/2[(k+r)^2C_{13}\log(k+r)], \qquad (76)\\
\theta_1 &= 1/9(k+r)^6(\eta_1+\mu)[4BrC_2^4(\eta_1-\mu)(-1+\xi^2)] \\
&\quad -1/(k+r)^4(\eta_1+\mu)[4BrC_2^3C_4(\eta_1-\mu)(-1+\xi^2)] \\
&\quad -BrC_4(k+r)^2(C_{13}+4C_4^3(-1+\xi^2)) \\
&\quad -1/(k+r)^2(\eta_1+\mu)[2BrC_2(C_{11}(\eta_1+\mu)+12C_4^2C_2\mu(-1+\xi^2))] \\
&\quad +A_{12} + A_{11}\log(k+r) \\
&\quad +2Br(C_{13}C_2 + 2C_4(C_{11}+8C_2C_4^2(-1+\xi^2)))\log(k+r)^2, \qquad (77)\\
\phi_1 &= -1/9(k+r)^6(\eta_1+\mu)[4BrC_2^4ScSr(\eta_1-\mu)(-1+\xi^2)] \\
&\quad +1/(k+r)^4(\eta_1+\mu)[4BrC_2^3C_4ScSr(\eta_1-\mu)(-1+\xi^2)] \\
&\quad +BrC_4(k+r)ScSr(C_{13}+4C_4^3(-1+\xi^2)) \\
&\quad +1/(k+r)^2(\eta_1+\mu)[2BrC_2ScSr(C_{11}(\eta_1+\mu)+12C_4^2C_2\mu(-1+\xi^2))] \\
&\quad +B_{12} + B_{11}\log(k+r) \\
&\quad -2BrScSr(C_{13}C_2 + 2C_4(C_{11}+8C_2C_4^2(-1+\xi^2)))\log(k+r)^2, \qquad (78)
\end{aligned}$$

$$\begin{aligned}
Z_1 &= \eta_x(\frac{A_{11}}{k+\eta} - Br\left(\frac{2C_2^2}{(k+\eta)^3} - 2C_4^2(k+\eta)\right) + \frac{8BrC_2^4(\eta_1-\mu)(\xi^2-1)}{3(k+\eta)^7(\eta_1+\mu)} \\
&\quad + \frac{16BrC_2^3C_4(\eta_1-\mu)(\xi^2-1)}{3(k+\eta)^5(\eta_1+\mu)} - 2BrC_4(k+\eta)\left(C_{13}+4C_4^3(\xi^2-1)\right) \\
&\quad + \frac{4BrC_2(C_{11}(\eta_1+\mu)+12C_2C_4^2\mu(\xi^2-1))}{(k+\eta)^3(\eta_1+\mu)} \\
&\quad + \frac{4Br(C_{13}C_2+2C_4(C_{11}+8C_2C_4^2(\xi^2-1))\ln(k+\eta))}{k+\eta}, \qquad (79)
\end{aligned}$$

in which

$$C_{11} = \frac{L_2(L_1+2C_2^2(-9C_4(k^2-\eta^2)^2(k^2+\eta^2)+C_2(3k^2+\eta^2)(k^2+3\eta^2)))}{(k-\eta)^4(k+\eta)^4},$$

$$C_{12} = \frac{L_2(L_3+C_2(9C_4(k^2-\eta^2)^2-4C_2(k^2+\eta^2)))}{(k^2-\eta^2)^4},$$

$$C_{13} = -4L_2C_4^3,$$

$$C_{14} = 0,$$

$$A_{11} = \frac{(-Br(Bi_2(N_1-\frac{N_2}{(k+\eta)^7(\eta_1+\mu)})+Bi_1(\frac{N_3}{(k-\eta)^7(\eta_1+\mu)}-N_4)))}{9N},$$

$$A_{12} = \frac{\begin{array}{c}Br(N_5+N_6+N_7-N_8-N_9+\frac{M(Bi_1(\frac{N_3}{(k-\eta)^7(\eta_1+\mu)}-N_4)+(Bi_2(N_1-\frac{N_2}{(k+\eta)^7(\eta_1+\mu)}))}{N}-N_9\\ -18Bi_1(C_{13}C_2+2C_4(C_{11}+8C_2C_4^2(-1+\xi^2)))\ln(k+\eta)^2)\end{array}}{9Bi_1},$$

$$B_{11} = \frac{-BrScSr(Bi_3(Z_2-\frac{Z_1}{(k-\eta)^7(\eta_1+\mu)})+Bi_4(\frac{Z}{(k+\eta)^7(\eta_1+\mu)}-Z_3))}{9Y},$$

$$B_{12} = \frac{BrScSr(N_7+N_8+N_9+Z_4-Z_5+Z_6-Z_7-\frac{36Bi_3C_2^3C_4(\eta_1-\mu)(-1+\xi^2)}{(k+\eta)^4(\eta_1+\mu)}+Z_8)}{9Bi_3}.$$

The constants appearing in these equations are written in Appendix A.

5. Discussion

The behavior of the axial velocity $u(y)$, temperature $\theta(y)$, concentration $\phi(y)$, and heat-transfer coefficient $Z(x)$ with respect to the influential parameters is described in this section.

5.1. Axial Velocity Distribution

Figure 2a–c examines the effect of various parameters on the axial velocity. Figure 2a clearly shows that the axial velocity increases with an increase in We. Such an increasing trend is due to increased relaxation time and viscosity decay. The effect of curvature parameter k on $u(y)$ is depicted in Figure 2b. It is observed that the axial velocity $u(y)$ decreases with an increase in the curvature k near the lower wall of the channel while the reverse situation is observed near the upper wall of the channel. Variation in $u(y)$ for the elastic parameters E_1, E_2, and E_3 are shown in Figure 2c. This Figure indicates that, by increasing E_3 (which represents the oscillatory resistance), the velocity $u(y)$ decreases and the axial velocity $u(y)$ increases by increasing $E_1$and E_2.

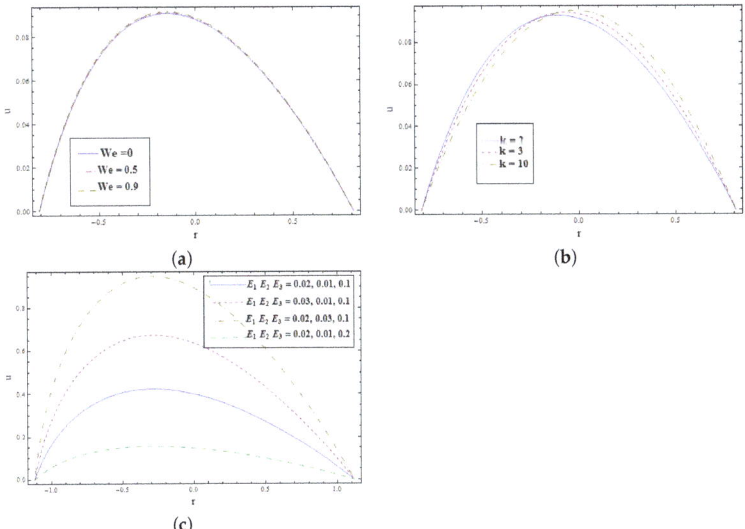

Figure 2. (**a**) Variation of We on u when $E_1 = 0.02$; $E_2 = 0.01$; $E_3 = 0.1$; $\epsilon = 0.2$; $k = 1.5$; $\xi = 0.5$; $\mu = 0.1$; $\eta_1 = 0.1$; $t = 0.1$; $x = -0.2$.; (**b**) Variation of k on u when $E_1 = 0.02$; $E_2 = 0.01$; $E_3 = 0.1$; $\epsilon = 0.2$; $We = 0.01$; $\xi = 0.5$; $\mu = 0.1$; $\eta_1 = 0.1$; $t = 0.1$; $x = -0.2$.; (**c**) Variation of E_1, E_2, E_3 on u when $\epsilon = 0.2$; $We = 0.2$; $k = 1.5$; $\xi = 0.5$; $\mu = 0.1$; $\eta_1 = 0.1$; $t = 0.1$; $x = 0.2$.

5.2. Temperature Distribution

Figure 3a–g indicates the influence of different parameters on the fluid temperature distribution $\theta(y)$. Figure 3a demonstrates that the magnitude of the temperature profile boosts while increasing the value of We as Weissenberg number is the ratio of ealstic forces and viscous forces, therefore, an increase in We dominates the viscosity and enhance the temperature of the fluid. It reveals that temperature is higher for Johnson–Segalman fluid than that of the viscous fluid temperature. Figure 3b reflects that when Brinkman number Br increases, the temperature goes up. This increase in the temperature is due to the viscous dissipation effects. Figure 3c portrays the effects of elastic parameters (E_1, E_2, and E_3) on the temperature profile $\theta(y)$. Increased temperature $\theta(y)$ can be seen with an increment in E_1 and E_2 and it decreases with increasing in E_3. Figure 3d depicts that the temperature falls drastically towards the lower portion of the channel and continues to rise in the upper portion of the channel as the curvature parameter k rises. Figure 3e portrays the slip parameter activity indicating that temperature declines as the slip parameter rises near the upper channel wall, while it has an opposite impact close to the lower boundary. Figure 3f illustrates that enhancing the Biot number Bi_1 reduces the temperature profile $\theta(y)$ near the upper inlet section but no impact has found in the lower inlet section. Similarly, Figure 3g reveals that the temperature profile $\theta(y)$ for Biot number Bi_2 declines near the lower inlet section and has no noticeable impact near the upper inlet section.

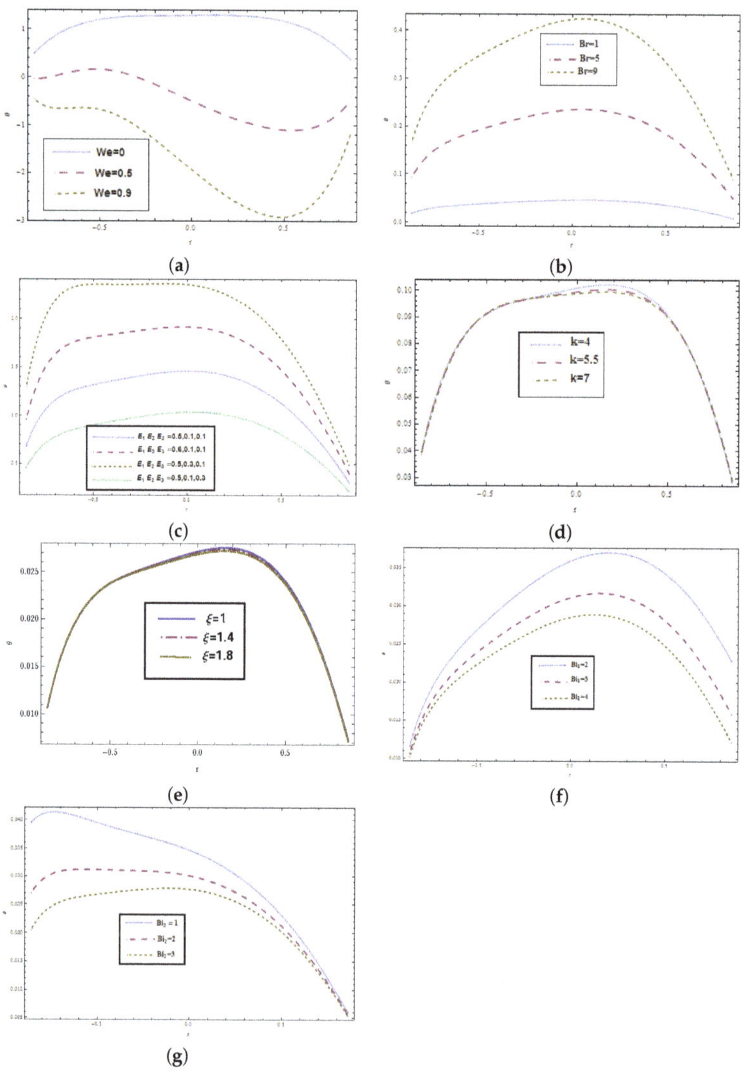

Figure 3. (**a**) effect of We on θ when $E_1 = 0.5$; $E_2 = 0.04$; $E_3 = 0.01$; $\epsilon = 0.15$; $k = 10$; $Br = 1.8$; $\xi = 1.8$; $\mu = 0.5$; $\eta_1 = 0.6$; $t = 0.1$; $x = -0.2$; $Bi_1 = 10$; $Bi_2 = 8$. (**b**) variation of Br on θ when $E_1 = 0.04$; $E_2 = 0.03$; $E_3 = 0.01$; $\epsilon = 0.15$; $k = 1.5$; $We = 0.01$; $\xi = 1.9$; $\mu = 0.6$; $\eta_1 = 0.8$; $t = 0.1$; $x = -0.2$; $Bi_1 = 10$; $Bi_2 = 8$. (**c**) variation of E_1, E_2, E_3 on θ when $\epsilon = 0.15$; $We = 0.01$; $k = 1.5$; $Br = 0.5$; $\xi = 1.9$;

$\mu = 0.5$; $\eta_1 = 0.8$; $t = 0.1$; $x = -0.2$; $Bi_1 = 10$; $Bi_2 = 8$. (**d**) variation of k on θ when $E_1 = 0.05$; $E_2 = 0.04$; $E_3 = 0.01$; $\epsilon = 0.15$; $We = 0.01$; $Br = 1.8$; $\xi = 2.9$; $\mu = 0.5$; $\eta_1 = 0.8$; $t = 0.1$; $x = -0.2$; $Bi_1 = 10$; $Bi_2 = 8$. (**e**) variation of ξ on θ when $E_1 = 0.05$; $E_2 = 0.03$; $E_3 = 0.01$; $\epsilon = 0.15$; $We = 0.01$; $Br = 1.5$; $k = 2.9$; $\mu = 0.6$; $\eta_1 = 0.8$; $t = 0.1$; $x = -0.2$; $Bi_1 = 10$; $Bi_2 = 8$. (**f**) Variation of Bi_1 on θ when $E_1 = 0.04$; $E_2 = 0.03$; $E_3 = 0.01$; $\epsilon = 0.15$; $We = 0.01$; $Br = 2.5$; $\xi = 1.9$; $\mu = 0.6$; $\eta_1 = 0.8$; $k = 1.5$ $t = 0.1$; $x = -0.2$; $Bi_2 = 8$. (**g**) variation of Bi_2 on θ when $E_1 = 0.04$; $E_2 = 0.03$; $E_3 = 0.01$; $\epsilon = 0.15$; $We = 0.01$; $Br = 2.5$; $\xi = 1.9$; $\mu = 0.6$; $\eta_1 = 0.8$; $k = 1.5$ $t = 0.1$; $x = -0.2$; $Bi_1 = 10$.

5.3. Concentration Distribution

Figure 4a–h represents the effects of emerging parameters on the fluid concentration distribution $\phi(y)$. Figure 4a depicts that concentration $\phi(y)$ increases when We increases due to increase in elasticity of the fluid as We physically represents the ratio of elastic to viscous forces. Figure 4b demonstrates that the concentration decreases when the Brinkman number intensifies. The effect of elastic parameters (E_1, E_2, and E_3) are represented in Figure 4c. Here, with the increase in E_1 and E_2, the concentration distribution decreases, while for E_3 concentration distribution $\phi(y)$ increases. Figure 4d shows the influence of slip parameter on $\phi(y)$. This Figure shows that the concentration decays in the lower half portion of the channel, while the reverse trend is seen throughout the upper half portion of the channel. Figure 4e shows that the fluid concentration reduces towards the upper wall of the channel and rises in the lower portion of the channel as the curvature parameter k rises. Figure 4f indicates that the concentration declines with an increase in the Schmidt number (Sc) which physically represents the ratio of momentum diffusivity and mass diffusivity. When we increase the value of Schmidt number, it actually dominates the mass diffusion and thus concentration of the fluid decays. The mass diffusion decays through increase in Schmidt number and, hence, concentration distribution $\phi(y)$ decreases. The effects of Biot numbers Bi_3 and Bi_4 are examined separately for the concentration profile $\phi(y)$ in the Figure 4g,h. It is found that variation of Bi_3 has significant effect near the upper wall and it hardly shows any effect near the lower wall. Similarly, the effects of Bi_4 are significant across the lower wall and the concentration profile $\phi(y)$ tends to decrease here. Biot number values are assumed to be greater than 1 and it indicates the non-uniform concentration fields inside the fluid. Also it reveals that convection is much quicker than conduction. From a realistic point of view, the parameters chosen are thus appropriate.

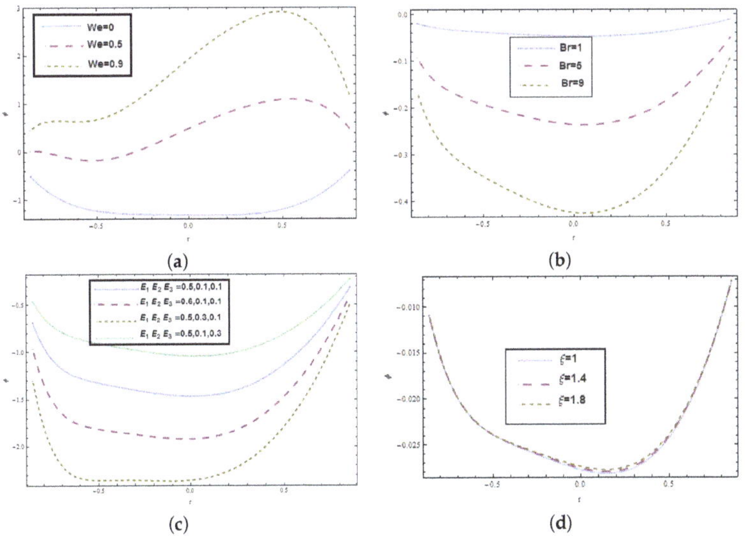

Figure 4. *Cont.*

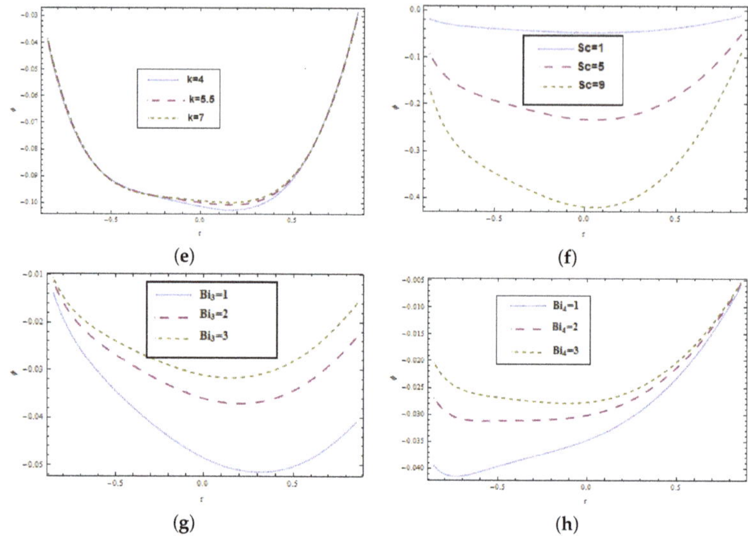

Figure 4. (**a**) Variation of We on ϕ when $E_1 = 0.5$; $E_2 = 0.04$; $E_3 = 0.01$; $\epsilon = 0.15$; $k = 10$; $Br = 0.5$; $\xi = 1.8$; $\mu = 0.5$; $\eta_1 = 0.6$; $t = 0.1$; $x = -0.2$; $Sc = 1$; $Sr = 1$; $Bi_1 = 10$; $Bi_2 = 8$. (**b**) Variation of Br on ϕ when $E_1 = 0.04$; $E_2 = 0.03$; $E_3 = 0.01$; $\epsilon = 0.15$; $k = 1.5$; $We = 0.01$; $\xi = 1.9$; $Sc = 1$; $Sr = 1$; $\mu = 0.6$; $\eta_1 = 0.8$; $t = 0.1$; $x = -0.2$; $Bi_1 = 10$; $Bi_2 = 8$. (**c**) Variation of E_1, E_2, E_3 on ϕ when $\epsilon = 0.15$; $We = 0.01$; $k = 1.5$; $Br = 0.5$; $\xi = 1.9$; $\mu = 0.5$; $\eta_1 = 0.8$; $t = 0.1$; $x = -0.2$; $Sc = 1$; $Sr = 1$; $Bi_1 = 10$; $Bi_2 = 8$. (**d**) Variation of ξ on ϕ when $E_1 = 0.05$; $E_2 = 0.03$; $E_3 = 0.01$; $\epsilon = 0.15$; $We = 0.01$; $Br = 0.5$; $k = 2.5$; $\mu = 0.6$; $\eta_1 = 0.8$; $t = 0.1$; $x = -0.2$; $Bi_1 = 10$; $Bi_2 = 8$; $Sc = 1$; $Sr = 1$. (**e**) variation of k on ϕ when $E_1 = 0.05$; $E_2 = 0.04$; $E_3 = 0.01$; $\epsilon = 0.15$; $We = 0.01$; $Br = 1.5$; $\xi = 2.9$; $\mu = 0.5$; $\eta_1 = 0.8$; $t = 0.1$; $x = -0.2$; $Bi_1 = 10$; $Bi_2 = 8$; $Sc = 1$; $Sr = 1$. (**f**) Variation of Sc on ϕ when $E_1 = 0.04$; $E_2 = 0.03$; $E_3 = 0.01$; $\epsilon = 0.15$; $We = 0.01$; $Br = 1$; $\xi = 1.9$; $k = 1.5$; $Sr = 1$; $\mu = 0.6$; $\eta_1 = 0.8$; $t = 0.1$; $x = -0.2$; $Bi_1 = 10$; $Bi_2 = 8$. (**g**) Variation of Bi_1 on ϕ when $E_1 = 0.04$; $E_2 = 0.03$; $E_3 = 0.01$; $\epsilon = 0.15$; $We = 0.01$; $Br = 0.5$; $\xi = 1.9$; $\mu = 0.6$; $\eta_1 = 0.8$; $k = 1.5$; $t = 0.1$; $x = -0.2$; $Bi_2 = 8$; $Sc = 1$; $Sr = 1$. (**h**) Variation of Bi_2 on ϕ when $E_1 = 0.04$; $E_2 = 0.03$; $E_3 = 0.01$; $\epsilon = 0.15$; $We = 0.01$; $Br = 0.5$; $\xi = 1.9$; $\mu = 0.6$; $\eta_1 = 0.8$; $k = 1.5$; $t = 0.1$; $x = -0.2$; $Bi_1 = 10$; $Sc = 1$; $Sr = 1$.

5.4. *Coefficient of Heat-Transfer*

In Figure 5a–e, we noticed the variability of the coefficient of heat transfer $Z(x)$ for We, Br, k, Bi_1, and Bi_2. Due to peristalsis, the nature of the heat transfer coefficient is oscillatory. The absolute value of the coefficient of overall heat transfer $Z(x)$ falls as We increases (see Figure 5a). Figure 5b illustrates that the coefficient of heat transfer results in an increase when Brinkman number intensifies. Figure 5c displays the curvature parameter's behavior. This indicates that the coefficient of heat transfer $Z(x)$ boosts with an increase in k. Further, Figure 5d,e analyze the effects of Biot numbers on heat transfer coefficient $Z(x)$. Increasing Bi_1 the magnitude of heat transfer coefficient $Z(x)$ increases and it decreases for Bi_2.

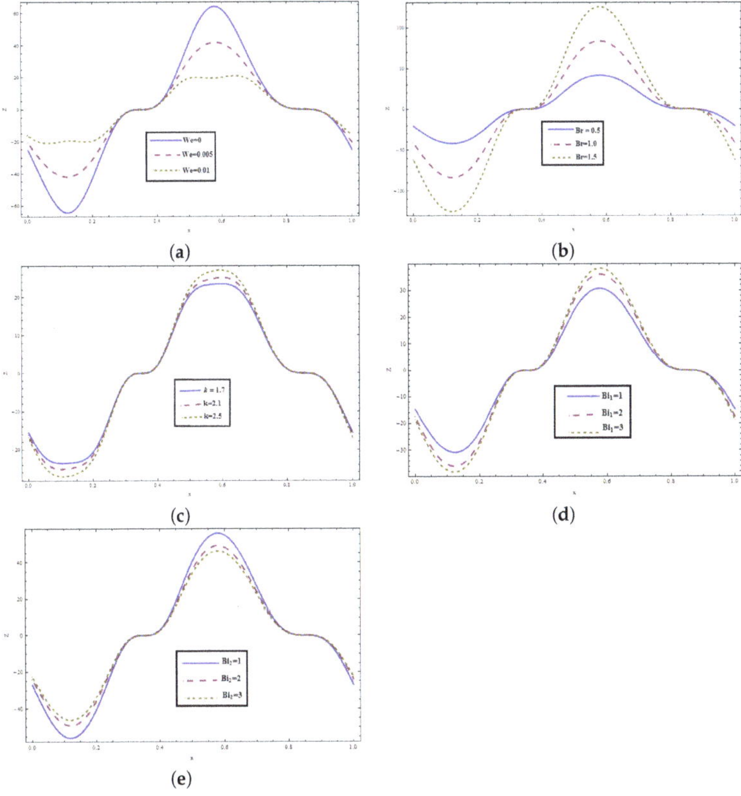

Figure 5. (**a**) Variation of We on Z when $E_1 = 0.5$; $E_2 = 0.04$; $E_3 = 0.01$; $\epsilon = 0.15$; $k = 10$; $Br = 0.5$; $\xi = 1.8$; $\mu = 0.5$; $\eta_1 = 0.6$; $t = 0.1$; $Bi_1 = 10$; $Bi_2 = 8$. (**b**) Variation of Br on Z when $E_1 = 0.5$; $E_2 = 0.04$; $E_3 = 0.01$; $\epsilon = 0.15$; $k = 10$; $We = 0.005$; $\xi = 1.8$; $\mu = 0.5$; $\eta_1 = 0.6$; $t = 0.1$; $Bi_1 = 10$; $Bi_2 = 8$. (**c**) Variation of k on Z when $E_1 = 0.5$; $E_2 = 0.04$; $E_3 = 0.01$; $We = 0.005$; $\epsilon = 0.15$; $k = 10$; $Br = 0.5$; $\xi = 1.8$; $\mu = 0.5$; $\eta_1 = 0.6$; $t = 0.1$; $Bi_1 = 10$; $Bi_2 = 8$. (**d**) Variation of Bi_1 on Z when $E_1 = 0.5$; $E_2 = 0.04$; $E_3 = 0.01$; $\epsilon = 0.15$; $We = 0.005$; $Br = 0.5$; $\xi = 1.8$; $\mu = 0.5$; $\eta_1 = 0.6$; $k = 10$; $t = 0.1$; $x = -0.2$; $Bi_2 = 8$. (**e**) Variation of Bi_2 on Z when $E_1 = 0.5$; $E_2 = 0.04$; $E_3 = 0.01$; $\epsilon = 0.15$; $We = 0.005$; $Br = 0.5$; $\xi = 1.8$; $\mu = 0.5$; $\eta_1 = 0.6$; $k = 10$; $t = 0.1$; $x = -0.2$; $Bi_1 = 10$.

6. Conclusive Remarks

This article addresses the peristalsis of Johnson-Segalman fluid in a circular channel with walls' compliance and convective heat and mass transfer conditions. Perturbation solution has been obtained under the long wave length and low Reynolds number approximation. The axial velocity of Johnson–Segalman is found to be greater than that of the Newtonian fluid. The velocity profile is skewed to the left because of curved channel, whereas the concentration and temperature profiles are inclined towards the right. Further, the velocity profile is not symmetric about the centre line in curved channel. At a certain level in the curved channel, the fluid approaches maximum velocity, which decreases in magnitude. The curved channel is transformed into the straight channel with relatively high value of curvature parameter. The results of this problem in Newtonian fluid model can be reduced when $m = \mu = 0$.

Author Contributions: Conceptualization, H.Y.; formal analysis, N.I.; investigation, A.H.; methodology, H.Y. and A.H.; software, A.H.; validation, H.Y. and N.I.; writing—original draft preparation, H.Y., N.I. and A.H.; writing—review and editing, N.I. and H.Y.; visualization, A.H. and N.I.; supervision, H.Y. All authors have read and agreed to the published version of the manuscript.

Funding: This research received no external funding.

Conflicts of Interest: The authors declare no conflict of interest.

Appendix A

The parameters appearing in the solutions are described here.

$$
\begin{aligned}
L_1 &= k(-8E_1\pi^3\epsilon\cos(2\pi(x-t)) - 8E_2\pi^3\epsilon\cos(2\pi(x-t)) + 4E_3\pi^2\epsilon\sin(2\pi(x-t)),\\
L_2 &= \frac{2\eta_1(-1+\xi^2)}{3(\eta_1+\mu)},\\
L_3 &= 3C_4^3(-(k-\eta)^2\ln(k-\eta) + (k+\eta)^2\ln(k+\eta)),\\
M &= \frac{1}{k+\eta} + Bi_1\ln(k+\eta),\\
M_1 &= \frac{1}{k-\eta} - Bi_2\ln(k+\eta),\\
M_2 &= \frac{Bi_1}{-k+\eta} + Bi_1Bi_2\ln(k-\eta),\\
M_3 &= -Bi_2M + M_2,\\
M_4 &= \frac{8C_2C_4\ln(k-\eta)}{(k-\eta)},\\
M_5 &= \frac{8C_2C_4\ln(k+\eta)}{(k+\eta)},\\
M_6 &= \frac{C_2^2(2+Bi_2(k-\eta))}{(k-\eta)^3},\\
M_7 &= \frac{C_2^2(-2+Bi_1(k+\eta))}{(k+\eta)^3},\\
M_8 &= (-2+Bi_2(k-\eta))(k-\eta),\\
M_9 &= (2+Bi_1(k+\eta))(k+\eta),\\
M_{10} &= 4Bi_1C_2C_4\ln(k+\eta)^2,\\
M_{11} &= 4Bi_2C_2C_4\ln(k-\eta)^2,\\
M_{12} &= \frac{C_2^2}{(k+\eta)} + C_4^2(k+\eta)^2,\\
M_{13} &= -\frac{2C_2^2}{(k+\eta)^3} + 2C_4^2(k+\eta),\\
M_{14} &= Bi_1(-BrM_1(M_{10}+M_5-M_7-C_4^2M_9) + BM(M_4+M_6+C_4^2M_8-M_{11}),\\
N &= \frac{Bi_1}{k-\eta} + \frac{Bi_2}{k+\eta} + Bi_1Bi_2(-\ln(k-\eta)+\ln(k+\eta)),\\
N_1 &= \frac{18(C_{13}C_2 + 2C_4(C_{11} + 8C_2C_4^2(-1+\xi^2)))\ln(k+\eta)(2+Bi_1(k+\eta))\ln(k+\eta)}{k+\eta},\\
N_2 &= (18(C_{11}C_2(k+\eta)^4(-2+Bi_1(k+\eta))(\eta_1+\mu) - 4C_2^4(-6+Bi_1(k+\eta))(\eta_1-\mu)(-1+\xi^2)\\
&\quad +36C_4C_2^3(-4+Bi_1(k+\eta))(\eta_1-\mu)(-1+\xi^2)\\
&\quad +216C_2^2C_4^2(k+\eta)^4(-2+Bi_1(k+\eta))\mu(-1+\xi^2)\\
&\quad +9C_4(k+\eta)^8(2+Bi_1(k+\eta))(\eta_1+\mu)(C_{13}+4C_4^3(-1+\xi^2))),
\end{aligned}
$$

$$
\begin{aligned}
N_3 &= (18(C_{11}C_2(k+\eta)^4(2+Bi_2(k-\eta))(\eta_1+\mu) - 4C_2^4(6+Bi_2(k-\eta))(\eta_1-\mu)(-1+\xi^2) \\
&\quad +36C_4C_2^3(4+Bi_2(k-\eta))(k-\eta)^2(\eta_1-\mu)(-1+\xi^2) \\
&\quad +216C_2^2C_4^2(k-\eta)^4(2+Bi_2(k-\eta))(k-\eta)^4\mu(-1+\xi^2) \\
&\quad +9C_4(k-\eta)^8(-2+Bi_2(k-\eta))(\eta_1+\mu)(C_{13}+4C_4^3(-1+\xi^2))), \\
N_4 &= \frac{18(C_{13}C_2+2C_4(C_{11}+8C_2C_4^2(-1+\xi^2)))\ln(k-\eta)(-2+Bi_2(k-\eta))\ln(k-\eta)}{k-\eta}, \\
N_5 &= \frac{18Bi_1C_2(C_{11}(\eta_1+\mu)+12C_2C_4^2\mu(-1+\xi^2))}{(k+\eta)^2(\eta_1+\mu)} + \frac{36Bi_1C_2^3C_4(\eta_1-\mu)(-1+\xi^2)}{(k+\eta)^4(\eta_1+\mu)}, \\
N_6 &= 18C_4(k+\eta)(C_{13}+4C_4^3(-1+\xi^2)) + 9Bi_1C_4(k+\eta)^2(C_{13}+4C_4^3(-1+\xi^2)) \\
&\quad +\frac{24C_2^2(\eta_1-\mu)(-1+\xi^2)}{(k+\eta)^7(\eta_1+\mu)}, \\
N_7 &= \frac{36(C_{13}C_2+2C_4(C_{11}+8C_2C_4^2(-1+\xi^2)))\ln(k+\eta)}{k+\eta}, \\
N_8 &= \frac{36C_2(C_{11}(\eta_1+\mu)+12C_2C_4^2\mu(-1+\xi^2))}{(k+\eta)^3(\eta_1+\mu)}, \\
N_9 &= \frac{144C_2^3C_4(\eta_1-\mu)(-1+\xi^2))}{(k+\eta)^5(\eta_1+\mu)}, \qquad N_{10} = \frac{4C_2^4(\eta_1-\mu)(-1+\xi^2))}{(k+\eta)^6(\eta_1+\mu)}, \\
Y &= \frac{Bi_3}{k-\eta} + \frac{Bi_4}{k+\eta} + Bi_3Bi_4(-\ln(k-\eta)+\ln(k+\eta)), \\
Y_1 &= \frac{C_2^2}{(k+\eta)^3} \quad C_4^2(k+\eta), \qquad Y_2 = \frac{2Bi_3\ln(k-\eta)}{k-\eta}, \\
Y_3 &= \frac{1}{(k-\eta)^3} + \frac{2Bi_4k\eta}{(k^2-\eta^2)^2}, \qquad Y_4 = Bi_3Bi_4\ln(k-\eta)^2, \\
Y_5 &= Bi_4\ln(k+\eta)(\frac{2}{k+\eta} + Bi_3\ln(k+\eta)), \\
Y_6 &= \frac{1}{(-k+\eta)^3} - \frac{2Bi_4k\eta}{(k^2-\eta^2)^2}, \\
Y_7 &= 4C_2C_4(Y_4 + \frac{2Bi_3\ln(k-\eta)}{-k+\eta} + Bi_4\ln(k+\eta)(\frac{-2}{k+\eta} - Bi_3\ln(k+\eta))), \\
Y_8 &= Y_7 + 2Bi_4(\frac{-C_2^2}{(k+\eta)^3} + C_4^2(k+\eta)) + 2Bi_3(C_2^2Y_6 + C_4^2(k-\eta+2Bi_4k\eta)), \\
Y_9 &= 4Bi_3C_2C_4\ln(k+\eta)^2 + \frac{Y_8(\frac{1}{k+\eta}+Bi_3\ln(k+\eta))}{Y}, \\
D &= (18(C_{11}C_2(k+\eta)^4(-2+Bi_3(k+\eta))(\eta_1+\mu) - 4C_2^4(-6+Bi_3(k+\eta))(\eta_1-\mu)(-1+\xi^2) \\
&\quad +36C_4C_2^3(k+\eta)^2(-4+Bi_3(k+\eta))(\eta_1-\mu)(-1+\xi^2) \\
&\quad +216C_2^2C_4^2(k+\eta)^4(-2+Bi_3(k+\eta))\mu(-1+\xi^2) \\
&\quad +9C_4(k+\eta)^8(2+Bi_3(k+\eta))(\eta_1+\mu)(C_{13}+4C_4^3(-1+\xi^2))), \\
D_1 &= (18(C_{11}C_2(k-\eta)^4(2+Bi_4(k-\eta))(\eta_1+\mu) - 4C_2^4(6+Bi_4(k-\eta))(\eta_1-\mu)(-1+\xi^2) \\
&\quad +36C_4C_2^3(k-\eta)^2(4+Bi_4(k-\eta))(\eta_1-\mu)(-1+\xi^2) \\
&\quad +216C_2^2C_4^2(k-\eta)^4(2+Bi_4(k-\eta))\mu(-1+\xi^2) \\
&\quad +9C_4(k-\eta)^8(-2+Bi_4(k-\eta))(\eta_1+\mu)(C_{13}+4C_4^3(-1+\xi^2))), \\
D_2 &= \frac{18(C_{13}C_2+2C_4(C_{11}+8C_2C_4^2(-1+\xi^2)))\ln(k-\eta)(-2+Bi_4(k-\eta))\ln(k-\eta)}{k-\eta}, \\
D_3 &= \frac{18(C_{13}C_2+2C_4(C_{11}+8C_2C_4^2(-1+\xi^2)))\ln(k+\eta)(2+Bi_3(k+\eta))\ln(k+\eta)}{k+\eta}, \\
D_4 &= 18Bi_3(C_{13}C_2+2C_4(C_{11}+8C_2C_4^2(-1+\xi^2)))\ln(k+\eta)^2,
\end{aligned}
$$

$$
\begin{aligned}
D_5 &= 18C_4(k+\eta)(C_{13}+4C_4^3(-1+\xi^2)+9Bi_3C_4(k+\eta)^2(C_{13}+4C_4^3(-1+\xi^2)) \\
&\quad +\frac{24C_2^2(\eta_1-\mu)(-1+\xi^2)}{(k+\eta)^5(\eta_1+\mu)}, \\
D_6 &= \frac{4Bi_3C_2^4(\eta_1-\mu)(-1+\xi^2)}{(k+\eta)^6(\eta_1+\mu)}, \\
D_7 &= \frac{18Bi_3C_2(C_{11}(\eta_1+\mu)+12C_2C_4^2\mu(-1+\xi^2))}{(k+\eta)^2(\eta_1+\mu)}, \\
D_8 &= \frac{1}{Y}(Bi_3(D_2-\frac{D_1}{(k-\eta)^7(\eta_1+\mu)})+Bi_4(\frac{D}{(k+\eta)^7(\eta_1+\mu)}-D_3))(\frac{1}{k+\eta}+Bi_3\ln(k+\eta)).
\end{aligned}
$$

References

1. Latham, T.W. *Fluid Motion in a Peristaltic Pump*; MIT: Cambridge, MA, USA, 1966.
2. Shapiro, A.H.; Jaffrin, M.Y.; Wienberg, S.L. Peristaltic pumping with long wavelengths at low Reynolds number. *J. Fluid Mech.* **1969**, *37*, 799–825. [CrossRef]
3. Yin, F.; Fung, Y.C. Peristaltic wave in circular cylindrical tubes. *J. Appl. Mech.* **1969**, *36*, 579–587. [CrossRef]
4. Hayat, T.; Iqbal, R.; Tanveer, A.; Alsaedi, A. Mixed convective peristaltic transport of Carreau-Yasuda nanofluid in a tapered asymmetric channel. *J. Mol. Liq.* **2016**, *223*, 1100–1113. [CrossRef]
5. Ramesh, K. Effects of slip and convective conditions on the peristaltic flow of couple stress fluid in an asymmetric channel through porous medium. *Comput. Meth. Prog. Biomed.* **2016**, *135*, 1–14. [CrossRef] [PubMed]
6. Tripathi, D.; Bég, O.A. Peristaltic propulsion of generalized Burgers' fluids through a non-uniform porous medium: A study of chyme dynamics through the diseased intestine. *Math. Biosci.* **2014**, *248*, 67–77. [CrossRef]
7. Shehzad, S.A.; Abbasi, F.M.; Hayat, T.; Alsaadi, F. Model and comparative study for peristaltic transport of water based nanofluids. *J. Mol. Liq.* **2015**, *209*, 723–728. [CrossRef]
8. Abbasi, F.M.; Hayat, T.; Ahmad, B. Peristaltic transport of copper–water nanofluid saturating porous medium. *Phys. Low Dimens. Nanostruct.* **2015**, *67*, 47–53. [CrossRef]
9. Gad, N.S. Effects of Hall currents on peristaltic transport with compliant walls. *Appl. Math. Comput.* **2014**, *235*, 546–554. [CrossRef]
10. Abbas, M.A.; Bai, Y.Q.; Bhatti, M.M.; Rashidi, M.M. Three dimensional peristaltic flow of hyperbolic tangent fluid in non-uniform channel having flexible walls. *Alex. Eng. J.* **2016**, *55*, 653–662. [CrossRef]
11. Lachiheb, M. Effect of coupled radial and axial variability of viscosity on the peristaltic transport of Newtonian fluid. *Appl. Math. Comput.* **2014**, *244*, 761–771. [CrossRef]
12. Shit, G.C.; Roy, M. Hydromagnetic effect on inclined peristaltic flow of a couple stress fluid. *Alex. Eng. J.* **2014**, *4*, 949–958. [CrossRef]
13. Ramesh, K.; Tripathi, D.; Bhatti, M.M.; Khalique, C.M. Electro-osmotic flow of hydromagnetic dusty viscoelastic fluids in a microchannel propagated by peristalsis. *J. Mol. Liq.* **2020**, *314*, 113568. [CrossRef]
14. Tariq, H.; Khan, A.A.; Zaman, A. Theoretical analysis of peristaltic viscous fluid with inhomogeneous dust particles. *Arab. J. Sci. Eng.* **2020**. [CrossRef]
15. Hina, S.; Mustafa, M.; Hayat, T.; Alotaibi, N.D. On peristaltic motion of pseudoplastic fluid in a curved channel with heat/mass transfer and wall properties. *Appl. Math. Comput.* **2015**, *263*, 378–391. [CrossRef]
16. Bhatti, M.M.; Elelamy, A.F.; Sait, S.M.; Ellahi, R. Hydrodynamics interactions of metachronal waves on particulate-liquid motion through a ciliated annulus: Application of bio-engineering in blood clotting and endoscopy. *Symmetry* **2020**, *14*, 532. [CrossRef]
17. Fomicheva, M.; Müller, W.H.; Vilchevskaya, E.N.; Bessonov, N. Funnel flow of a Navier-Stokes-fluid with potential applications to micropolar media. *Facta Univ. Ser. Mech. Eng.* **2019**, *17*, 255–267. [CrossRef]
18. Javed, M. A mathematical framework for peristaltic mechanism of non-Newtonian fluid in an elastic heated channel with Hall effect. *Multidiscip. Model. Mater. Struct.* **2020**. [CrossRef]
19. Bhandari, D.S.; Tripathi, D.; Narla, V.K. Pumping flow model for couple stress fluids with a propagative membrane contraction. *Int. J. Mech. Sci.* **2020**. [CrossRef]
20. Hayat, T.; Wang, Y.; Siddiqi, A.M.; Hutter, K. Peristaltic flow of a Johnson-Segalman fluid in a planar channel. *Math. Probl. Eng.* **2003**, *2003*, 1–23. [CrossRef]

21. Hayat, T.; Mumtaz, S. Resonant oscillations of plate in an electrically conducting rotating Johnson-Segalman fluid. *Comp. Math. Appl.* **2005**, *50*, 1669–1676. [CrossRef]
22. Hayat, T.; Afsar, A.; Ali, N. Peristaltic transport of a Johnson–Segalman fluid in an asymmetric channel. *Math. Comput. Model.* **2005**, *47*, 380–400. [CrossRef]
23. Elshahed, M.; Haroun, M.H. Peristaltic transport of Johnson–Segalman fluid under effect of a magnetic field. *Math. Probl. Eng.* **2005**, *2005*, 663–677. [CrossRef]
24. Wang, Y.; Hayat, T.; Hutter, K. Peristaltic transport of a Johnson–Segalman fluid through a deformable tube. *Theor. Comput. Fluid Dyn.* **2007**, *21*, 369–380. [CrossRef]
25. Sato, H.; Kawai, T.; Fujita, T.; Okabe, M. Two dimensional peristaltic flow in curved channels. *Trans. Jpn. Soc. Mech. Eng. B* **2000**, *66*, 679–685. [CrossRef]
26. Ali, N.; Sajid, M.; Hayat, T. Long wavelength flow analysis in a curved channel. *Z. Naturforsch.* **2010**, *65*, 191–196. [CrossRef]
27. Majeed, A.; Javed, T.; Ghaffari, A.; Rashidi, M.M. Analysis of heat transfer due to stretching cylinder with partial slip and prescribed heat flux: A Chebyshev Spectral Newton Iterative Scheme. *Alex. Eng. J.* **2015**, *54*, 1029–1036. [CrossRef]
28. Hina, S.; Mustafa, M.; Hayat, T.; Alsaedi, A. Peristaltic flow of Powell-Eyring fluid in curved channel with heat transfer: A useful application in biomedicine. *Comp. Meth. Prog. Biomed.* **2016**, *135*, 89–100. [CrossRef]
29. Abbasi, F.M.; Hayat, T.; Alsaedi, A. Numerical analysis for MHD peristaltic transport of Carreau-Yasuda fluid in a curved channel with Hall effects. *J. Magn. Magn. Mat.* **2015**, *382*, 104–110. [CrossRef]
30. Ali, N.; Javid, K.; Sajid, M.; Zaman, A.; Hayat, T. Numerical simulations of Oldroyd 8-constant fluid flow and heat transfer in a curved channel. *Int. J. Heat Mass Transf.* **2016**, *94*, 500–508. [CrossRef]
31. Ahmed, R.; Ali, N.; Khan, S.U.; Tlili, I. Numerical simulations for mixed convective hydromagnetic peristaltic flow in a curved channel with joule heating features. *AIP Adv.* **2020**, *10*, 075303. [CrossRef]
32. Hina, S.; Hayat, T.; Alsaedi, A. Heat and mass transfer effects on the peristaltic flow of Johnson–Segalman fluid in a curved channel with compliant walls. *Int. J. Heat Mass Transf.* **2012**, *55*, 3511–3521. [CrossRef]
33. Mehmood, O.U.; Qureshi, A.A.; Yasmin, H.; Uddin, S. Thermo-mechanical analysis of non Newtonian peristaltic mechanism: Modified heat flux model. *Phys. A Stat. Mech. Appl.* **2020**, *550*, 124014. [CrossRef]
34. Hayat, T.; Javid, M.; Hendi, A.A. Peristaltic transport of viscous fluid in a curved channel with compliant walls. *Int. J. Heat Mass Transf.* **2011**, *54*, 1615–1621. [CrossRef]
35. Riaz, A.; Ellahi, R.; Sait, S.M. Role of hybrid nanoparticles in thermal performance of peristaltic flow of Eyring—Powell fluid model. *J. Them. Anal. Calorim.* **2020**. [CrossRef]
36. Hayat, T.; Yasmin, H.; Al-Yami, M. Soret and Dufour effects in peristaltic transport of physiological fluid with chemical reaction: A mathematical analysis. *Comput. Fluids* **2014**, *89*, 242–253. [CrossRef]
37. Yasmin, H.; Farooq, S.; Awais, M.; Alsaedi, A.; Hayat, T. Significance of heat and mass process in peristalsis of a rheological material. *Heat Trans. Res.* **2019**, *50*, 1561–1580. [CrossRef]
38. Hayat, T.; Yasmin, H.; Ahmad, B.; Chen, B. Simultaneous effects of convective conditions and nanoparticles on peristaltic motion. *J. Mol. Liq.* **2014**, *193*, 74–82. [CrossRef]
39. Hayat, T.; Bibi, A.; Yasmin, H.; Alsaadi, F.E. Magnetic field and thermal radiation effects in peristaltic flow with heat and mass convection. *J. Ther. Sci. Eng. Appl.* **2018**, *10*, 051018. [CrossRef]
40. Yasmin, H.; Hayat, T.; Alsaedi, A.; Alsulami, H.H. Peristaltic transport of Johnson-Segalman fluid in an asymmetric channel with convective boundary conditions. *Appl. Math. Mech.-Engl. Ed.* **2014**, *35*, 697–716. [CrossRef]

Article

Hydrodynamics Interactions of Metachronal Waves on Particulate-Liquid Motion through a Ciliated Annulus: Application of Bio-Engineering in Blood Clotting and Endoscopy

Muhammad Mubashir Bhatti [1], Asmaa F. Elelamy [2], Sadiq M. Sait [3] and Rahmat Ellahi [4,5,6,*]

1 College of Mathematics and Systems Science, Shandong University of Science and Technology, Qingdao 266590, Shandong, China; mmbhatti@sdust.edu.cn
2 Department of Mathematics, Faculty of Education, Ain Shams University, El Makrizy Street, Roxy, Heliopolis, 11566 Cairo, Egypt; asmaafathe@edu.asu.edu.eg
3 Center for Communications and IT Research, Research Institute, King Fahd University of Petroleum & Minerals, Dhahran 31261, Saudi Arabia; sadiq@kfupm.edu.sa
4 Department of Mathematics & Statistics, FBAS, IIUI, Islamabad 44000, Pakistan
5 Fulbright Fellow Department of Mechanical Engineering, University of California, Riverside, CA 92521, USA
6 Center for Modeling & Computer Simulation, Research Institute, King Fahd University of Petroleum and Minerals, Dhahran 31261, Saudi Arabia
* Correspondence: rellahi@alumni.ucr.edu

Received: 17 February 2020; Accepted: 26 March 2020; Published: 3 April 2020

Abstract: This study deals with the mass transport phenomena on the particle-fluid motion through an annulus. The non-Newtonian fluid propagates through a ciliated annulus in the presence of two phenomenon, namely (i) endoscopy, and (ii) blood clot. The outer tube is ciliated. To examine the flow behavior we consider the bi-viscosity fluid model. The mathematical modeling has been formulated for small Reynolds number to examine the inertia free flow. The purpose of this assumption is that wavelength-to-diameter is maximal, and the pressure could be considerably uniform throughout the entire cross-section. The resulting equations are analytically solved, and exact solutions are given for particle- and fluid-phase profiles. Computational software Mathematica has been used to evaluate both the closed-form and numerical results. The graphical behavior across each parameter has been discussed in detail and presented with graphs. The trapping mechanism is also shown across each parameter. It is noticed clearly that particle volume fraction and the blood clot reveal converse behavior on fluid velocity; however, the velocity of the fluid reduced significantly when the fluid behaves as a Newtonian fluid. Schmidt and Soret numbers enhance the concentration mechanism. Furthermore, more pressure is required to pass the fluid when the blood clot appears.

Keywords: cilia motion; blood clot; endoscopy; mass transport; particle-fluid

---

1. Introduction

Flagella and cilia are two distinct names, but are used interchangeably for similar structure of eukaryotic cells. In animals, cilia, which are hair-like appendages, are prominent in the digestive system, respiratory system, reproductive tracts of human beings, and the nervous systems. The movement of cilia plays an essential part in physiological systems, i.e., circulation, respiration, locomotion, alimentation, spermatic fluid propagation, reproduction, etc. It is well-known that ciliary and flagellar movements consist of active sliding, similar to the peristaltic flow of fluid in smooth muscles, whereas flagellar is more complicated. Cilia can be split into two categories, i.e., non-motile and motile. When cilia and flagella are close to each other, they manifest a propagation of waves on a large scale known as *Metachronal waves*. Beating cilia produce metachronal waves over the surface in large numbers, on the ciliated surface of protozoa, and the adjoining activity of cilia coordinates via the hydrodynamics interactions. It is worth mentioning here that metachronal waves are self-organized. Cilia produce bending waves to derive single cells through a medium or to push the fluid over the surface of a fixed cell. The standard form of the ciliary model, motivated by the cilium structure or axoneme, is known as sliding filament model. As shown in Figure 1, the axoneme structure has nine microtubule doublets around the outside, and two are located in the center. Most of the cilia beat at approximately 10–40 Hz, but the form of beating varies. Their length starts from 2 μm to several millimeters, and their diameters are approximatley 0.2 μm. As a result, a low Reynolds number (Re→0) approximation can be applied.

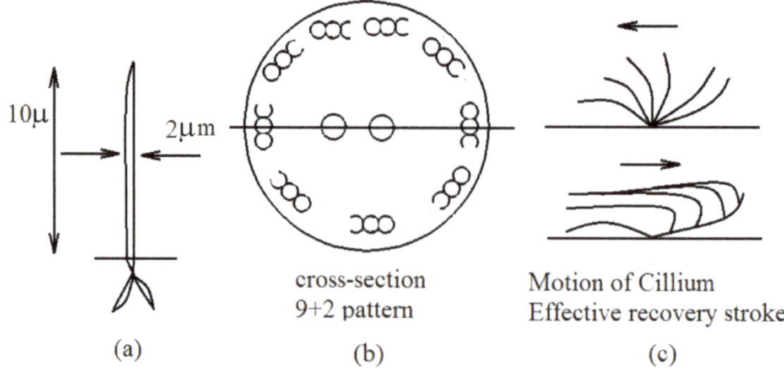

Figure 1. Strutcure and size of cilia [1] (**a**) typical dimension, (**b**) cross-section, (**c**) cilia stroking.

Nadeem et al. [2] considered the Carreau fluid model to examine the cilia motion through a symmetric channel using the perturbation method. Nadeem and Sadaf [3] discussed the cilia motion of viscous nanofluid through the curvy compliant channel. They used a homotopy analysis method to examine the closed-form solution against the temperature and velocity profile. Maiti and Pandey [4] presented a theoretical study on the nonlinear cilia motion using the Power-law fluid model. Abo-Elkhair et al. [5] used the Adomian decomposition scheme to simluate the cilia motion of magneto-fluid through a ciliated channel. Bhatti et al. [6] discussed the impact of the magnetic field on a ciliated channel using the particle-fluid mechanism. Ashraf et al. [7] examined the peristaltic cilia motion through a human fallopian tube using a Newtonian fluid model. And finally, Ramesh et al. [8] used the behavior of magnetized couple stress fluid model moving through a ciliated channel

Particles in fluid appear in multifarious applications, including biology, geology, chemical engineering, and fluid mechanics [9] to name a few. Several industrial processes include fluidized catalyst beds, pneumatic propagation, and sedimentation. Further, in the biological systems, it involves the flow of blood in the cardiovascular system. The collisions among the particles and the fluids may influence the rheological and the viscosity behavior of the suspension. Particle-wall and particle-particle interactions produce the migration of particles, which causes the anisotropic particle micro-structures and clusters [10,11]. At the mesoscopic scale, a well-known example of the particle-fluid interaction is the movement of the red blood cells (RBC). The flow behavior of the RBC plays a pivotal role in the different pathological and physiological mechanisms. For instance, the rotation and random transverse propagation of RBC in a shear flow plays an essential role in thrombogenesis [12]. These types of movements are firmly associated with the interaction of RBC to RBC and fluid (i.e., plasma) to RBC since one RBC is obstructed by another coming towards it from below or above. RBC is the essential determinant of the blood characteristics in micro-circulation due to their large volume fraction in blood and their aggregability. Mekheimer and Abd Elmaboud [13] investigated the peristaltic motion of fluid having solid particles through different forms of annulus and showed the exact solutions. Mekheimer and Mohamed [14] presented an application of a clot blood model using particle-fluid flow through an annulus. Further, they considered the pulsatile flow and obtained analytical solutions. Bhatti et al. [15] discussed the behavior of slip effects using a non-Newtonian fluid model that contains spherical particles. Bhatti and Zeeshan [16] explained the blood flow through an annulus filled with particles and fluid in the presence of a variable magnetic field. Some critical analysis of multiphase simulations are given in the following references [17–21].

Mass transfer with heat on Particle-fluid through fixed and fluidized beds play an essential role and provide necessary information for the development and design of numerous mass and heat transfer operation and chemical reactors including a system of particle and fluid. Gireesha et al. [22] investigated the particle-fluid suspension mechanism through a non-isothermal stretching plate in the presence of a magnetic field and radiative heat flux. They applied a numerical method to obtain the solutions. Bhatti et al. [23] presented a mathematical model of particle-fluid motion induced by a peristaltic wave with thermal radiation and electromagnetohydrodynamics effects. Kumar et al. [24] considered a particle-fluid motion with a nonlinear Williamson fluid model towards a stretching sheet with heat transfer effects. Bhatti et al. [25] explored the particle-fluid motion with heat and mass transfer using Sisko fluid model through a Darcy–Brinkman–Forchheimer porous medium. Some relevant studies on particle-fluid with mass and heat transfer are given in the references [26–28].

The main goal of the present study is to examine the mass transport on the particle-fluid suspension through a ciliated annulus with endoscopy and blood clot effects. Endoscopy plays an essential role in exploring the problems in human organs. In the mentioned studies, mostly work has been done with endoscopy and blood clot with simple Newtonian and non-Newtonian fluid models. In contrast, the present study deals with mass transport on particle-fluid motion through a ciliated annulus under different effects. Cilia motion plays a critical part, i.e., ciliary imperfections tend to create several human diseases. A genetic change compromises an appropriate function of the ciliopathies, cilia, which results in chronic disorders, i.e., primary ciliary dyskinesia and Senior–Loken syndrome or nephronophthisis. Furthermore, a flaw in primary cilium in renal tubule cells causes polycystic kidney disease. Ectopic pregnancy can occur due to a lack of functional cilia in a fallopian tube. If the cilia fail to move, then a fertilized ovum is unable to reach the uterus, which results in the ovum implant in a fallopian tube and tubal pregnancy will occur, which is the most usual type of ectopic pregnancy [29]. Therefore, the present study is essential to fill this gap and also beneficial to overcome the difficulties. Bi-viscosity fluid model is considered to examine the flow behavior. The governing mathematical modeling is performed under low Reynolds number approximation. Exact solutions are given for the fluid- and particulate-phase. The physical action of all the leading parameters is discussed against velocity, concentration, temperature profile, and the trapping mechanism is also presented through streamlines.

2. Problem Formulation

Consider two-dimensional co-axial infinite tubes. The outer tube is ciliated. The cylindrical coordinate system is selected, i.e., $\tilde{r}$ lies toward the radial direction, and $\tilde{z}$ lies toward the middle of an inner and the outer tube as given in Figure 2. The inner area between both the tubes is filled with bi-viscosity fluid. The flow is irrotational and the fluid is incompressible having constant viscosity. The fluid contains small spherical particles. The stress tensor for bi-viscosity fluid model [30] is defined as:

$$\chi = \begin{cases} 2\left[y_s/\sqrt{2\pi} + \tilde{\mu}_B\right]\zeta_{ij}, & \pi \geq \pi_{\Upsilon}, \\ 2\left[y_s/\sqrt{2\pi} + \tilde{\mu}_B\right]\zeta_{ij}, & \pi \leq \pi_{\Upsilon}, \end{cases} \tag{1}$$

where $\tilde{\mu}_B$ the plastic viscosity, Υ the volume fraction density, ζ_{ij} the deformation rate of component, and y_s the yield stress, π denotes the second invariant tensor of ζ_{ij}, π_{Υ} represents the critical value comprises on the non-Newtonian fluid model.

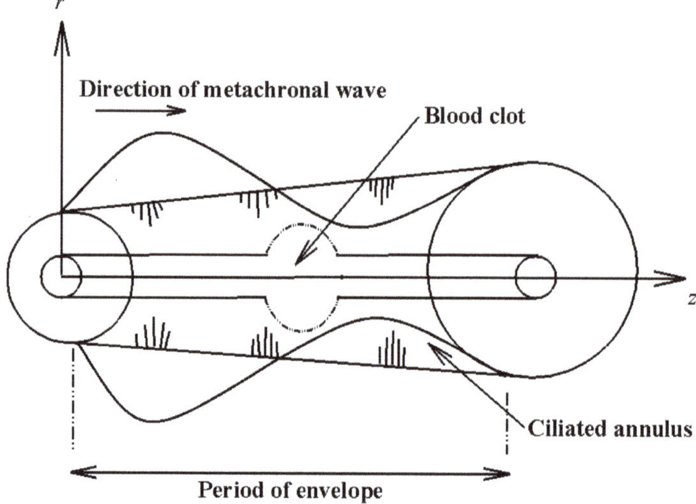

Figure 2. Blood flow structure through an ciliated annulus.

The Mathematical expression for the envelope of the cilia tips reads as [31,32]:

$$\tilde{r} = \hbar_1 = f_1(\tilde{t}, \tilde{z}) = b_0 + b_0\phi \cos\left[k(\tilde{z} - c\tilde{t})\right], \tag{2}$$

$$\tilde{z} = \hbar_2 = g_1(\tilde{t}, \tilde{z}) = \tilde{z}_0 + b_0\alpha\phi \sin\left[k(\tilde{z} - c\tilde{t})\right], \tag{3}$$

where $k = 2\pi/\lambda$, z_0 describes the reference location of the cilia, the non-dimensional parameter ϕ, which combines with b_0 (mean radius of the outer tube) in the form of $b_0\phi$ and represents the amplitude of metachronal wave, λ is the metachronal wavelength, c the velocity, and α describes the measure of the eccentricity of the elliptical motion.

The vertical and axial velocities are evaluated as [31,32]:

$$\tilde{u} = \frac{\partial \tilde{r}}{\partial \tilde{t}} = \frac{\partial f_1}{\partial \tilde{t}} + \frac{\partial f_1}{\partial \tilde{z}}\frac{\partial \tilde{z}}{\partial \tilde{t}} = \frac{\partial f_1}{\partial \tilde{t}} + \frac{\partial f_1}{\partial \tilde{z}}\tilde{u}, \;\; \tilde{z} = \tilde{z}_0, \tag{4}$$

$$\tilde{v} = \frac{\partial \tilde{z}}{\partial \tilde{t}} = \frac{\partial g_1}{\partial \tilde{t}} + \frac{\partial g_1}{\partial \tilde{z}}\frac{\partial \tilde{z}}{\partial \tilde{t}} = \frac{\partial g_1}{\partial \tilde{t}} + \frac{\partial g_1}{\partial \tilde{z}}\tilde{u}, \;\; \tilde{z} = \tilde{z}_0, \tag{5}$$

After some mathematical manipulation, Equations (4) and (5) read as:

$$\tilde{u} = -\frac{b_0\alpha c k\phi \cos\left[k(\tilde{z} - c\tilde{t})\right]}{1 - b_0\alpha k\phi \cos\left[k(\tilde{z} - c\tilde{t})\right]}, \tag{6}$$

$$\tilde{v} = \frac{b_0 c k\phi \sin\left[k(\tilde{z} - c\tilde{t})\right]}{1 - b_0\alpha k\phi \cos\left[k(\tilde{z} - c\tilde{t})\right]}. \tag{7}$$

The above boundary conditions help us to discriminate between the effective stroke and less effective recovery stroke of the cilia by considering the shortening of the cilia.

In view of above frame work, the mathematical modeling for the fluid- and particulate-phase is as follows [33]:

(i) *Fluid phase*

The continuity and momentum equations are proposed as:

$$\varphi\frac{\partial\tilde{v}_f}{\partial\tilde{r}}+\varphi\frac{\tilde{v}_f}{\tilde{r}}+\varphi\frac{\partial\tilde{u}_f}{\partial\tilde{z}}=0, \tag{8}$$

$$\varphi\frac{\partial\tilde{p}}{\partial\tilde{r}}-CS(\tilde{v}_p-\tilde{v}_f)=\varphi\left[\frac{1}{\tilde{r}}\frac{\partial}{\partial\tilde{r}}r\chi_{\tilde{r}\tilde{r}}+\frac{\partial}{\partial\tilde{z}}\chi_{\tilde{r}\tilde{z}}-\frac{\chi_{\tilde{\theta}\tilde{\theta}}}{\tilde{r}}\right], \tag{9}$$

$$\varphi\frac{\partial\tilde{p}}{\partial\tilde{z}}-CS(\tilde{u}_p-\tilde{u}_f)=\varphi\left[\frac{\partial}{\partial\tilde{z}}\chi_{\tilde{z}\tilde{z}}+\frac{1}{\tilde{r}}\frac{\partial}{\partial\tilde{r}}\tilde{r}\chi_{\tilde{r}\tilde{z}}\right], \tag{10}$$

The energy equation for the current flow is described as

$$\varphi\rho_f\tilde{c}\left[\frac{\partial}{\partial\tilde{t}}+\mathbf{V}_f\cdot\nabla\right]T_f=\kappa\varphi\nabla^2T_f+\varphi\chi_{\tilde{r}\tilde{z}}\left[\frac{\partial u_f}{\partial\tilde{r}}\right]+\frac{\rho_pc_pC}{\omega_T}(T_p-T_f), \tag{11}$$

The concentration equation for the current flow is described as

$$\varphi\left[\frac{\partial}{\partial\tilde{t}}+\mathbf{V}_f\cdot\nabla\right]K_f=D_m\varphi\nabla^2K_f+\frac{\rho_pC}{\rho_f\omega_c}(\varphi_p-\varphi_f)+\frac{D_m}{T_m}\varphi K_T\nabla^2T_f. \tag{12}$$

where $\varphi=1-C$.

(ii) *Particulate phase*

The continuity and momentum equation for this case read as

$$C\frac{\partial\tilde{v}_p}{\partial\tilde{r}}+C\frac{\tilde{v}_p}{\tilde{r}}+C\frac{\partial\tilde{u}_p}{\partial\tilde{z}}=0, \tag{13}$$

$$C\frac{\partial\tilde{p}}{\partial\tilde{r}}-SC(\tilde{v}_f-\tilde{v}_p)=0, \tag{14}$$

$$C\frac{\partial\tilde{p}}{\partial\tilde{z}}-SC(\tilde{u}_f-\tilde{u}_p)=0, \tag{15}$$

In this case, the energy equation is described as

$$\rho_pCc_p\left[\frac{\partial}{\partial\tilde{t}}+\mathbf{V}_p\cdot\nabla\right]T_p=\frac{\rho_pc_pC}{\omega_T}(T_f-T_p), \tag{16}$$

The concentration equation is described as

$$\left[\frac{\partial}{\partial\tilde{t}}+\mathbf{V}_p\cdot\nabla\right]K_p=\frac{1}{\omega_c}(K_f-K_p), \tag{17}$$

where S the drag coefficient, ρ the density of the fluid, C the particle volume fraction density, T the temperature, ω_T the thermal equilibrium time, ω_c is the required time period by a particle to regulate its concentration associated to the fluid, D_m the mass diffusivity coefficient, K_T is the thermal diffusion ratio, T_m the mean temperature, κ the thermal conductivity, c_p particle-phase specific heat, and $\tilde{c}$ the specific heat at constant volume.

The mathematical form of drag coefficient is expressed as [23]

$$S = \frac{9}{2}\frac{\mu_0}{B_0^2}\Phi(C), \Phi(C) = \frac{3C + 4 + [C(8-3C)]^{1/2}}{(2-3C)^2}, \tag{18}$$

where B_0 is the radius of each particle, and μ_o the fluid viscosity. The empirical relation for the viscosity suspension is expressed as [23]

$$\tilde{m} = 0.70e^{\left[\frac{1107}{T}e^{(-1.69C)}+2.49C\right]}, \mu_s = \frac{\mu_0}{1-\tilde{m}C}, \tag{19}$$

where μ_s denotes the viscosity of particle fluid mixture.

It is noted here that the results reduced for dusty-gas flows for small particle volume fraction as presented by Marble [34].

Defining the following non-dimensional variables

$$\begin{aligned} r &= \frac{\tilde{r}}{b_0}, u_{f,p} = \frac{\tilde{u}_{f,p}}{c}, z = \frac{\tilde{z}}{\lambda}, v_{f,p} = \frac{\lambda\tilde{v}_{f,p}}{b_0 c}, t = \frac{\tilde{t}c}{\lambda}, p = \frac{b_0^2}{\lambda\mu_0 c}\tilde{p}, \bar{\mu} = \frac{\mu_s}{\mu_0}, \\ v_0 &= \frac{\tilde{v}_0}{c}, r_1 = \frac{\tilde{r}_1}{b_0}, r_2 = \frac{\tilde{r}_2}{b_0}, \theta_{f,p} = \frac{T_{f,p} - T_0}{T_1 - T_0}, \vartheta_{f,p} = \frac{K_{f,p} - K_0}{K_1 - K_0}. \end{aligned} \tag{20}$$

Applying Equation (20) in Equations (8)–(18), and applying the approximation of low Reynolds number and ignoring the inertial forces. The resulting equations are found as

$$0 = \frac{\partial p}{\partial r}, \tag{21}$$

$$\frac{\partial p}{\partial z} = \frac{CSb_0^2}{\varphi\mu_0}\left(u_p - u_f\right) + \frac{\bar{\mu}}{r}\eta\frac{\partial}{\partial r}\left(r\frac{\partial u_f}{\partial r}\right). \tag{22}$$

It is noted here that the results for Newtonian fluid model can be recovered by taking $\zeta \to \infty$. The temperature and concentration equations read as

$$\frac{1}{r}\frac{\partial}{\partial r}\left(r\frac{\partial\theta_f}{\partial r}\right) + B_n\bar{\mu}\eta\left(\frac{\partial u_f}{\partial r}\right)^2 = 0, \tag{23}$$

$$\frac{1}{r}\frac{\partial}{\partial r}\left(r\frac{\partial\vartheta_f}{\partial r}\right) + S_cS_r\frac{1}{r}\frac{\partial}{\partial r}\left(r\frac{\partial\theta_f}{\partial r}\right) = 0, \tag{24}$$

where $\eta = (1 + 1/\zeta)$, B_n the Brinkman number, δ defines the wave number, S_c the Schmidt number, Pr the Prandtl number, S_r the Soret number, Ec the Eckert number, and ζ the fluid parameter. These parameters are defined as

$$\begin{aligned} B_n &= \mathrm{EcPr}, \mathrm{Pr} = \frac{\tilde{c}v\rho_f}{\kappa}, \zeta = \frac{\tilde{\mu}_B\sqrt{2\pi_Y}}{y_s}, \mathrm{Ec} = \frac{c^2}{\tilde{c}(T_1 - T_0)}, S_c = \frac{\nu}{D_m}, \delta = \frac{b_0}{\lambda}, \\ S_r &= \frac{D_m K_T}{\nu T_m}\left(\frac{\theta_1 - \theta_0}{\vartheta_1 - \vartheta_0}\right). \end{aligned} \tag{25}$$

The particulate-phase equations are found as

$$\frac{\partial p}{\partial z} + \frac{Sb_0^2}{\mu_0}\left(u_p - u_f\right) = 0, \tag{26}$$

$$\theta_f = \theta_p, \tag{27}$$

$$\vartheta_f = \vartheta_p. \tag{28}$$

From Equation (21), it is found that p cannot be the function of r. The relevant boundary conditions read as

$$u_f(r_1) = v_0, \theta_f(r_1) = 1, \qquad r_1 = a(z) = a_0 + h_c e^{-\pi^2(z-z_d-0.5)^2}, \tag{29}$$

$$u_f(r_2) = -\frac{2\pi\alpha\phi\delta\cos 2\pi\Xi}{1-2\pi\alpha\phi\delta\cos 2\pi\Xi}, \theta_f(r_2) = 0, \qquad r_2 = 1 + \frac{\lambda\Gamma z}{b_0} + \phi\sin 2\pi\Xi, \tag{30}$$

where $\Xi = (z-t)$, Γ is a constant which represents the magnitude that relies on the annulus length and its exit inlet dimensions, maximum height of the clot denoted by h_c, v_0 typify the velocity of the inner tube, the axial displacement of the clot is denoted by z_d, and the radius of the inner tube which makes the clot in the appropriate place is denoted by a_0. The results for endoscopy can be reduced by considering $h_c = 0$ in Equation (31) as a particular case of the present study.

3. Solutions of the Proposed Problem

Equations (22)–(24) are solved analytically using a computational software "Mathematica 10.3v", and the exact solutions are presented below:

$$u_f = C_1 + rC_2 + C_3 r\log r, \tag{31}$$

$$u_p = C_1 + rC_2 + C_3 r\log r - \frac{\mu_0}{Sb_0^2}\frac{dp}{dz}, \tag{32}$$

The solutions for the temperature profile for particulate- and fluid-phase are found as

$$\theta_{f,p} = \theta_0 + r^2\theta_1 + r^4\theta_2 + \theta_3\log r + \theta_4\log^2 r, \tag{33}$$

The solutions for the concentration profile for particulate- and fluid-phase are found as

$$\vartheta_{f,p} = \vartheta_0 + r^2\vartheta_1 + r^4\vartheta_2 + \vartheta_3\log r + \vartheta_4\log^2 r + \vartheta_5\log^3 r, \tag{34}$$

and the constants appearing in above Equations (31)–(34) i.e. $C_n, \theta_n, \vartheta_n$ $(n = 1,2\ldots)$ are given Appendix A.

The instantaneous volume flow rate for the present flow read as

$$Q(t,z) = 2\pi\varphi\int_{r_1}^{r_2} ru_f\mathrm{d}r + 2\pi C\int_{r_1}^{r_2} ru_p\mathrm{d}r. \tag{35}$$

The pressure gradient can be obtained after solving the above equation.

The pressure rise along the whole ciliated annulus can be determined as

$$\Delta p = \int_0^{L/\lambda} \wp\mathrm{d}z, \tag{36}$$

where $\wp$ represents the pressure gradient.

4. Graphical Analysis

In thissection, using the obtained numerical results we analyze the behavior of all the physical parameter. Particularly, we determine the behavior of velocity profile, concentration, and the temperature profile, against the height of the clot h_c, particle volume fraction C, wave number δ, Soret number S_r, Brinkmann number B_n, Schmidt number S_c, and eccentricity of the elliptic path of cilia α. Following parametric values [1] are chosen to analyze the graphical performance of all the leading parameters, i.e., $b_0 = 1.25$ cm, $\phi = 0.1 - 0.5$, $C = 0 - 0.6$, $\alpha = 0.3 - 1$, $\Gamma = 3b_0/\lambda$, $L = \lambda = 8.01$ cm , $\delta = 0.05 - 0.2$. Furthermore, the results for single-phase model can be recovered by considering $C = 0$ in the governing equations (see Equations (21)–(28)). Assume that the instantaneous volume flow rate is periodic in Ξ, i.e.,

$$\frac{Q}{\pi} = -\frac{\phi^2}{2} + \frac{\bar{Q}}{\pi} + 2\phi \sin 2\pi\Xi + \frac{2\phi\lambda z}{b_0}\Gamma \sin 2\pi\Xi + \phi^2 \sin^2 2\pi\Xi, \tag{37}$$

where $\bar{Q}$ denotes the average time flow rate over one period of wavelength.

Figure 3 depicts the behavior of blood clots and particle volume fraction on the velocity profile. We can observe from this figure that an increment in particle volume fraction C significantly suppresses the velocity profile. The velocity profile shows a decreasing behavior for endoscopic case, i.e., $h_c = 0$, whereas it increases due to the blood clot $h_c = 0.15$. It can be observed from Figure 4 that both parameters α and δ cause a positive impact on the velocity profile while its trend becomes reverse when $r > 1.35$. Figure 5 shows a plot of velocity profile against numerous values of ϕ. It can be seen from this figure that the velocity profile is remarkably suppressed with increments in ϕ. Furthermore, we also noticed that as compared with non-Newtonian case $\zeta = 0.1$, the fluid velocity lessen more when the fluid behaves as a Newtonian model $\zeta \to \infty$.

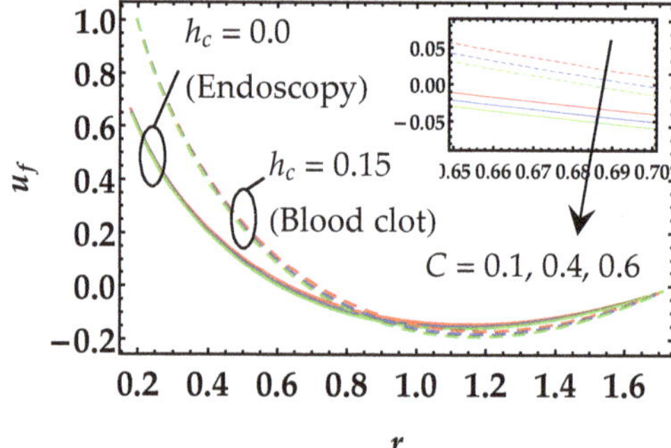

Figure 3. Velocity curves for different values of h_c and C.

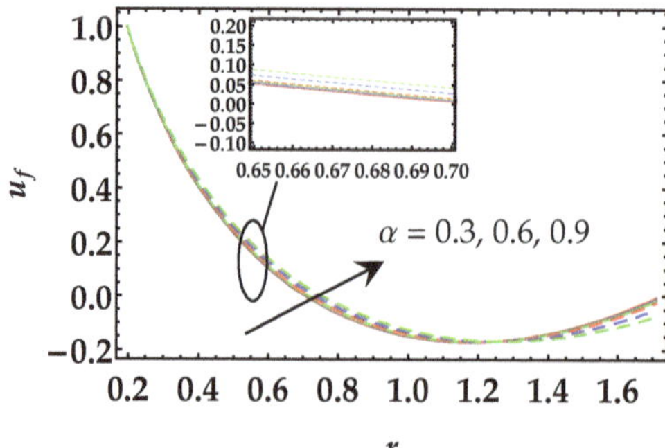

Figure 4. Velocity curves for different values of α and δ. Solid line: $\delta = 0.05$, dashed line: $\delta = 0.2$.

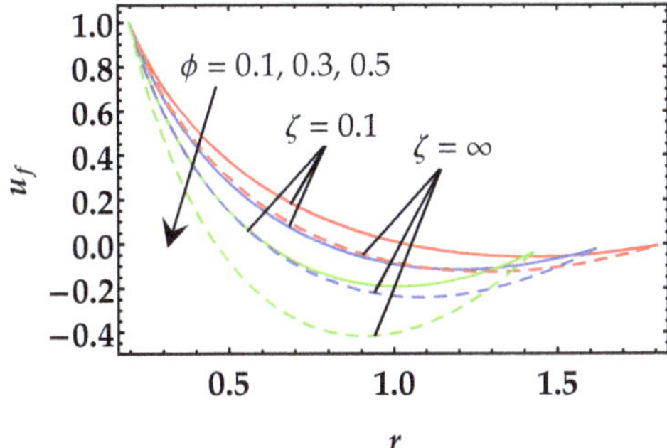

Figure 5. Velocity curves for different values of ϕ and ζ.

In Figures 6–8 we see the mechanism of temperature profile plotted against the multiple leading parameters. From Figure 6, we can see that the temperature profile rises due to the increment in particle volume fraction C. Further, we noticed that for the blood clot case $h_c = 0.15$, the temperature profile is increasing and has a higher magnitude as compared with the endoscopic case $h_c = 0$. It is analyzed from Figure 7 that the parameters α and δ restrain the temperature profile. Unfortunately, both parameters have small effects, especially when the wavenumber is very small at $\delta = 0.05$. Figure 8 shows plots with multiple values of Brinkman number B_n. Brinkman number represents the product of Eckert and Prandtl number Ec $\times$ Pr. Generally, it is the ratio between heat generated due to viscous dissipation and transport of heat due to molecular conduction. It can noticed that the temperature profile remarkably increases for higher values of Brinkman number. However, a similar behavior is observed against the higher values of ϕ.

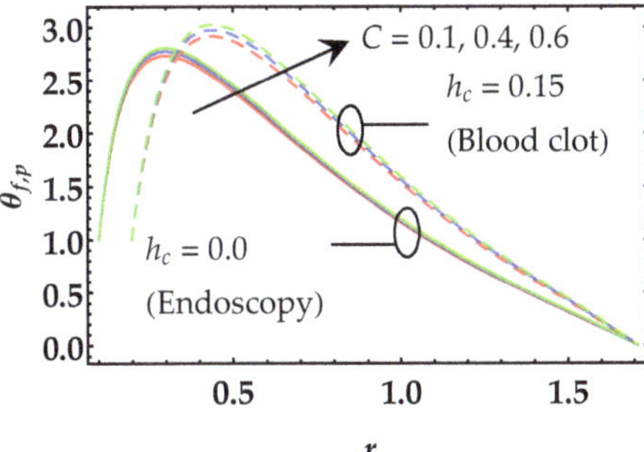

Figure 6. Temperature distribution for different values of h_c and C.

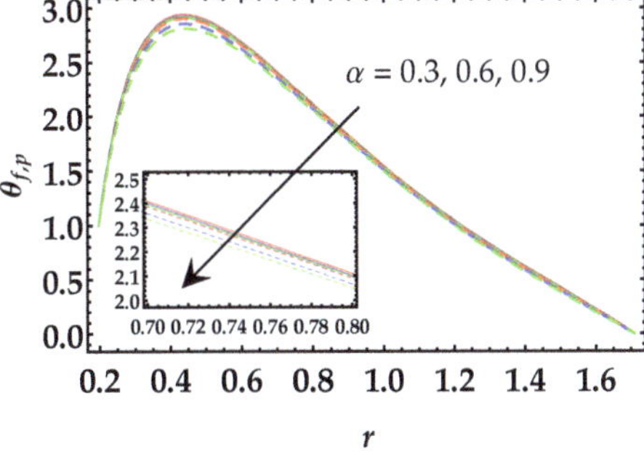

Figure 7. Temperature distribution for different values of α and δ. Solid line: $\delta = 0.05$, dashed line: $\delta = 0.2$.

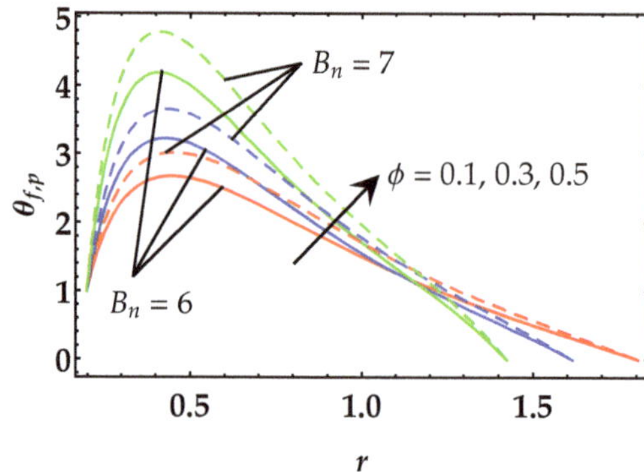

Figure 8. Temperature distribution for different values of ϕ and B_n.

Figure 9 is illustrated to analyze the mechanisms of Schmidt number S_c and Soret number S_r on the concentration profile. We can see from this figure that the concentration profile shows a decreasing mechanism against both parameters and remains uniform throughout the entire region. An increment in Schmidt number indicates that the viscous diffusion rate is more dominant as compared with the molecular diffusion rate, whose results tend to decline the concentration profile. Similarly, when the Soret number increases, the Thermophoresis forces generated, which oppose the concentration profile.

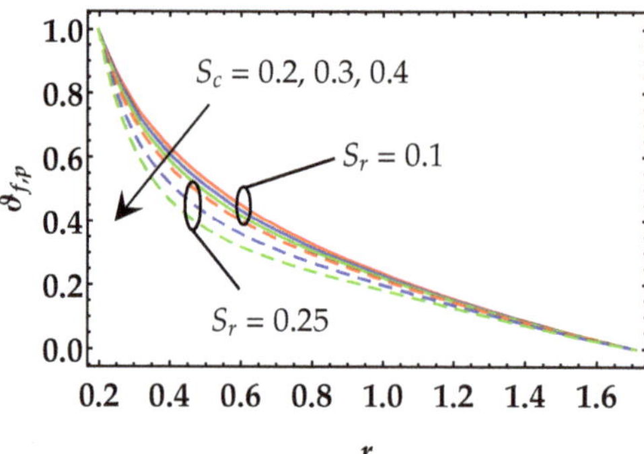

Figure 9. Concentration distribution for different values of S_c and S_r.

Figures 10–12 depict the variation of pressure rise versus time against different values of emerging parameters. It can be observed from Figure 10 that by enhancing the particle volume fraction C, the pressure rise is significantly decreasing, while due to the presence of blood clot, more pressure is required to pass the fluid. Further, we can see that the pressure rise is maximum in the region when $t \in (0.3, 0.7)$. It is clear from the Figure 11 that both parameters α and δ reveal versatile behavior on the pressure rise. We can also see that there are two critical points, for instance, at $t = 0.4$ and $t = 0.9$. In the region $t \in (0.4, 0.9)$ the pressure rise acts as an increasing function whereas in the other area it decreases. Similarly, we can observe that the pressure rise increases due to the increment in ϕ, as shown in Figure 12.

Trapping mechanism is presented in Figures 13–15 for different values of α, δ and h_c. It can be noticed from Figure 13 that by increasing the values of α, the trapping bolus reduces, and a number of boluses disappear. Similarly, in Figure 14, we can see that the higher values of δ tend to diminish the immensity of the trapping bolus, whereas the number of trapping bolus increase and streamlines increases. It is seen in Figure 15 that when the height of the blood clot increases, then streamlines shrink , and trapping bolus increase significantly.

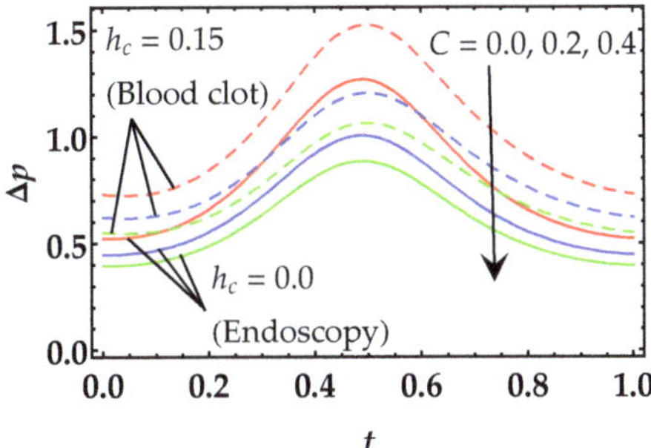

Figure 10. Pressure rise for different values of h_c and C.

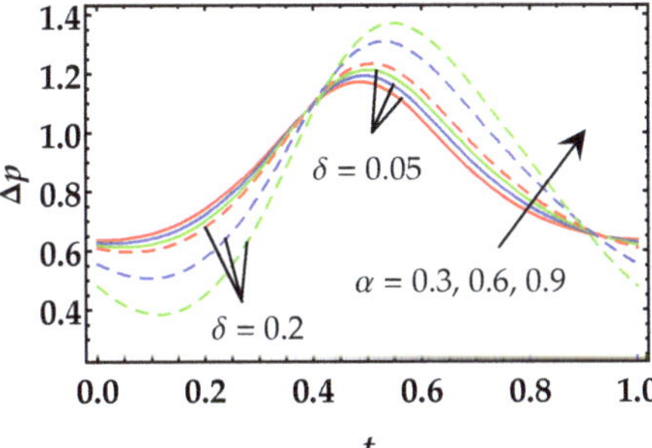

Figure 11. Pressure rise for different values of α and δ.

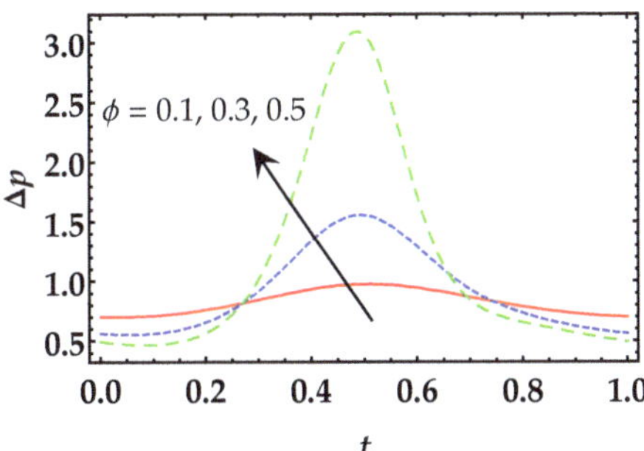

Figure 12. Pressure rise for different values of ϕ.

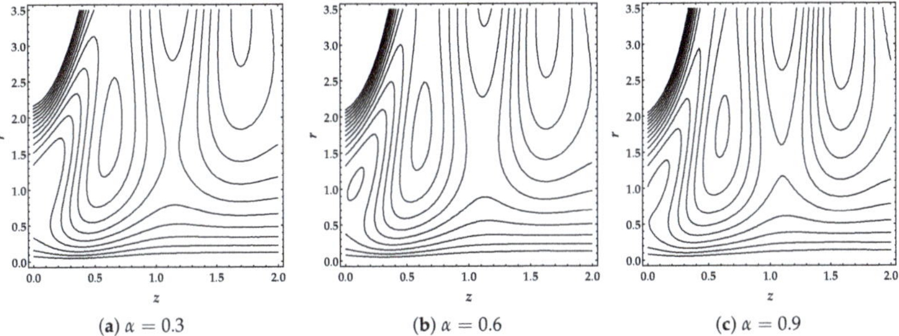

Figure 13. Trapping mechanism for different values of α.

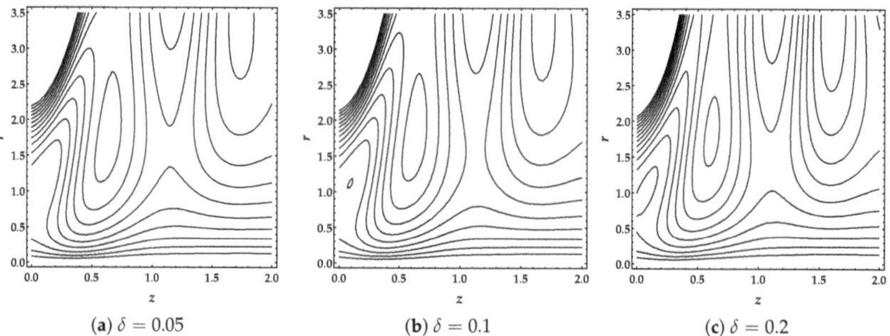

(a) $\delta = 0.05$ (b) $\delta = 0.1$ (c) $\delta = 0.2$

Figure 14. Trapping mechanism for different values of δ.

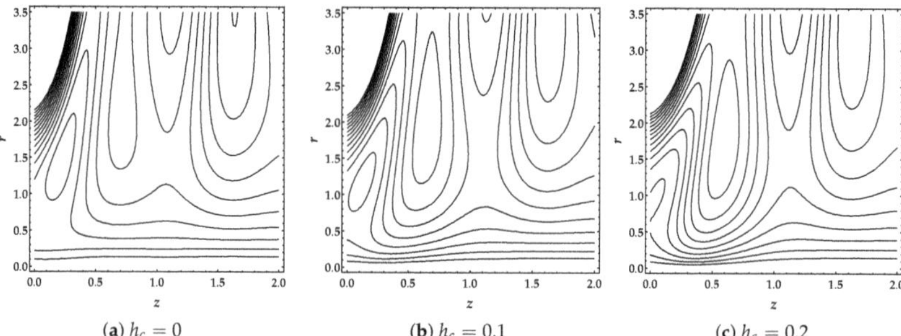

(a) $h_c = 0$ (b) $h_c = 0.1$ (c) $h_c = 0.2$

Figure 15. Trapping mechanism for different values of h_c.

5. Conclusions

In this study, we explained the behavior of particle-fluid with mass and heat transfer through a ciliated annulus. The effects of endoscopy and blood clot are also taken into account. To analyze the behavior of fluid in an annulus, we considered the bi-viscosity fluid model. The mathematical formulation is undertaken for low Reynolds number approximation. The formulated differential equations are analytically solved, and closed-form solutions are presented. The main observations of the present study are followed as:

(i) It is noticed that particle volume strongly opposes the flow, whereas the fluid velocity also decreases for the endoscopic case as compared with the blot clot case.
(ii) Velocity of the fluid also rises due to the enhancement in α and δ.
(iii) Temperature profile shows a higher magnitude in the presence of solid particles, while similar behavior is noticed during blood clot.
(iv) Brinkman number shows the dominant behavior on the temperature profile and enhances the temperature profile remarkably.
(v) Concentration profile reveals a decreasing behavior with the increase in the values of Soret and Schmidt numbers.
(vi) α and δ depict versatile behavior on the pressure rise profile.
(vii) Particle volume fraction opposes the pressure rise, whereas the blood clot enhances the pressure rise.
(viii) Trapping mechanism shows that the number of bolus gets bigger, and the streamlines gather as the height of the blood clot increases.

Furthermore, in this study, several effects have been ignored, i.e., magnetic field, porosity, chemical reaction and activation energy, respectively, which can be considered in future research.

Author Contributions: Investigation, M.M.B.; Methodology & Conceptualization, R.E.; Validation, A.F.E.; Writing—review & editing, S.M.S. All authors have read and agreed to the published version of the manuscript.

Acknowledgments: Authors thanks to those who are working in front line to save the humanity from corona pandemic in China and across the globe. We appreciate their devotion and services.

Funding: This research received no external funding.

Conflicts of Interest: The authors declare no conflict of interest.

Appendix A

$$C_1 = \frac{1}{4\varphi\eta}\frac{dp}{dz}, \tag{A1}$$

$$C_2 = \frac{1}{4\varphi\eta}\left[\frac{dp}{dz}(r_2^2 - r_1^2) - 4\xi\varphi\eta\right],, \tag{A2}$$

$$C_3 = \frac{\theta_5}{4\varphi\eta\zeta\log\frac{r_1}{r_2}},, \tag{A3}$$

$$\xi = -\frac{2\pi\alpha\varepsilon\delta\cos 2\pi\Xi}{1 - 2\pi\alpha\varepsilon\delta\cos 2\pi\Xi}. \tag{A4}$$

$$\theta_0 = \frac{1}{16\theta_7^2\log\frac{r_1}{r_2}}\Big[-\{16\theta_2^2 + B_n\theta_6 r_1^2(8\theta_5 + \theta_6 r_1^2)\eta\} - 8B_n\theta_5^2\eta\log r_1^2\log r_2 + B_n\eta\log r_1\left(\theta_6 r_2^2(8\theta_5 + \theta_6 r_2^2) + 8\theta_5^2\log^2 r_2\right)\Big], \tag{A5}$$

$$\theta_1 = -\frac{8B_n\theta_5\theta_6\eta}{16\theta_7^2}, \tag{A6}$$

$$\theta_2 = -\frac{B_n\theta_6^2\eta}{16\theta_7^2}, \tag{A7}$$

$$\theta_3 = \frac{1}{16\theta_7^2\log\frac{r_1}{r_2}}\left[16\theta_7^2 + B_n\theta_6(r_1^2 - r_2^2)\{8\theta_5 + \theta_6(r_1^2 + r_2^2)\}\eta + 8B_n\theta_5^2\eta\log^2\frac{r_1}{r_2}\right], \tag{A8}$$

$$\theta_4 = -\frac{B_n\theta_5^2\eta}{2\theta_7^2}, \tag{A9}$$

$$\theta_5 = \zeta\left(\frac{dp}{dz}(r_1^2 - r_2^2) + (\xi - v_0)4\varphi\eta\right), \tag{A10}$$

$$\theta_6 = 2\zeta\frac{dp}{dz}\log\frac{r_2}{r_1}, \tag{A11}$$

$$\theta_7 = 4\varphi\eta\zeta \log \frac{r_2}{r_1}. \tag{A12}$$

$$\begin{aligned}\vartheta_0 = & \frac{1}{192\theta_7^2 \log \frac{r_1}{r_2}} \Big[3(-64\theta_7^2 + B_n\theta_6 r_1^2(16\theta_5 + 3\theta_6 r_1^2)S_c S_r \eta) \log r_2 \\ & + 96B_n\theta_5^2 S_c S_r \eta \log^2 r_1 \log r_2 + 16B_n\theta_5^2 S_c S_r \eta \log r_1^3 \log r_2 \\ & + S_c S_r \log r_1 \{-3B_n\theta_5 r_2^2(16\theta_5 + 3\theta_6 r_2^2)\eta + 2\log r_2(48\theta_7^2 + 3B_n \\ & \times\theta_6(r_1^2 - r_2^2)(8\theta_5 + \theta_6(r_1^2 + r_2^2))\eta - 8B_n\theta_5^2\eta \log r_2(6 + \log r_2))\}\Big],\end{aligned} \tag{A13}$$

$$\vartheta_1 = \frac{B_n S_c S_r \theta_5 \theta_6 \eta}{\theta_7^2}, \tag{A14}$$

$$\vartheta_2 = \frac{3B_n S_c S_r \theta_6^2 \eta}{64\theta_7^2}, \tag{A15}$$

$$\begin{aligned}\vartheta_3 = & \frac{1}{192\theta_7^2 \log \frac{r_1}{r_2}} \Big[192\theta_7^2 - 3B_n\theta_5(r_1^2 - r_2^2)(16\theta_5 + 3\theta_6(r_1^2 + r_2^2))S_c S_r \eta \\ & + 2S_c S_r\{\log r_1(-48\theta_7^2\zeta - 3B_n\theta_6(r_1^2 - r_2^2)(8\theta_5 + \theta_6(r_1^2 + r_2^2))\eta - 8B_n \\ & \theta_5^2\eta \log r_1(6 + \log r_1))3\zeta(16\theta_7^2 + B_n\theta_6(r_1^2 - r_2^2)(8\theta_5 + \theta_6(r_1^2 + r_2^2))\eta \\ & + 8B_n\theta_5^2\eta \log r_1^2) \log r_2 + 24B_n\theta_5^2\eta(2 + \log r_1) \log r_2^2 + 8B_n\theta_5^2\, \eta \log r_2^3\}\Big],\end{aligned} \tag{A16}$$

$$\begin{aligned}\vartheta_4 = & \frac{1}{32\theta_7^2 \log \frac{r_1}{r_2}} \Big[S_c S_r\{16\theta_7^2 + B_n\theta_6(r_1^2 - r_2^2)(8\theta_5 + \theta_6(r_1^2 + r_2^2))\eta \\ & + 8B_n\theta_5^2\eta \log \frac{r_1}{r_2}(2 + \log r_1 + \log r_2)\}\Big],\end{aligned} \tag{A17}$$

$$\vartheta_5 = -\frac{B_n\theta_5^2 S_c S_r}{6\theta_7^2}\eta. \tag{A18}$$

References

1. Lardner, T.; Shack, W. Cilia transport. *Bull. Math. Biophys.* **1972**, *34*, 325–335. [CrossRef] [PubMed]
2. Nadeem, S.; Munim, A.; Shaheen, A.; Hussain, S. Physiological flow of Carreau fluid due to ciliary motion. *AIP Adv.* **2016**, *6*, 035125. [CrossRef]
3. Nadeem, S.; Sadaf, H. Ciliary motion phenomenon of viscous nanofluid in a curved channel with wall properties. *Eur. Phys. J. Plus* **2016**, *131*, 65. [CrossRef]
4. Maiti, S.; Pandey, S. Rheological fluid motion in tube by metachronal waves of cilia. *Appl. Math. Mech.* **2017**, *38*, 393–410. [CrossRef]
5. Abo-Elkhair, R.; Mekheimer, K.S.; Moawad, A. Cilia walls influence on peristaltically induced motion of magneto-fluid through a porous medium at moderate Reynolds number: Numerical study. *J. Egypt. Math. Soc.* **2017**, *25*, 238–251. [CrossRef]
6. Bhatti, M.; Zeeshan, A.; Rashidi, M. Influence of magnetohydrodynamics on metachronal wave of particle-fluid suspension due to cilia motion. *Eng. Sci. Technol. Int. J.* **2017**, *20*, 265–271. [CrossRef]

7. Ashraf, H.; Siddiqui, A.M.; Rana, M.A. Fallopian tube assessment of the peristaltic-ciliary flow of a linearly viscous fluid in a finite narrow tube. *Appl. Math. Mech.* **2018**, *39*, 437–454. doi:10.1007/s10483-018-2305-9. [CrossRef]
8. Ramesh, K.; Tripathi, D.; Bég, O.A. Cilia-assisted hydromagnetic pumping of biorheological couple stress fluids. *Propuls. Power Res.* **2019**, *8*, 221–233. [CrossRef]
9. Soo, S.L. *Particulates And Continuum-Multiphase Fluid Dynamics: Multiphase Fluid Dynamics*; Routledge: Abingdon, UK, 2018.
10. Hu, H.H.; Joseph, D.D.; Crochet, M.J. Direct simulation of fluid particle motions. *Theor. Comput. Fluid Dyn.* **1992**, *3*, 285–306. [CrossRef]
11. Feng, J.; Huang, P.; Joseph, D. Dynamic simulation of sedimentation of solid particles in an Oldroyd-B fluid. *J. Non-Newton. Fluid Mech.* **1996**, *63*, 63–88. [CrossRef]
12. Lima, R.; Ishikawa, T.; Imai, Y.; Takeda, M.; Wada, S.; Yamaguchi, T. Radial dispersion of red blood cells in blood flowing through glass capillaries: the role of hematocrit and geometry. *J. Biomech.* **2008**, *41*, 2188–2196. [CrossRef] [PubMed]
13. Mekheimer, K.S.; Abd Elmaboud, Y. Peristaltic transport of a particle–fluid suspension through a uniform and non-uniform annulus. *Appl. Bion. Biomech.* **2008**, *5*, 47–57. [CrossRef]
14. Mekheimer, K.S.; Mohamed, M.S. Peristaltic transport of a pulsatile flow for a particle-fluid suspension through a annular region: Application of a clot blood model. *Int. J. Sci. Eng. Res.* **2014**, *5*, 849–859.
15. Bhatti, M.; Zeeshan, A.; Ijaz, N. Slip effects and endoscopy analysis on blood flow of particle-fluid suspension induced by peristaltic wave. *J. Mol. Liquids* **2016**, *218*, 240–245. [CrossRef]
16. Bhatti, M.M.; Zeeshan, A. Study of variable magnetic field and endoscope on peristaltic blood flow of particle-fluid suspension through an annulus. *Biomed. Eng. Lett.* **2016**, *6*, 242–249. [CrossRef]
17. Maskeen, M.M.; Mehmood, O.U.; Zeeshan, A. Hydromagnetic solid–liquid pulsatile flow through concentric cylinders in a porous medium. *J. Visual.* **2018**, *21*, 407–419. [CrossRef]
18. Ellahi, R.; Zeeshan, A.; Hussain, F.; Abbas, T. Two-phase couette flow of couple stress fluid with temperature dependent viscosity thermally affected by magnetized moving surface. *Symmetry* **2019**, *11*, 647. [CrossRef]
19. Zeeshan, A.; Hussain, F.; Ellahi, R.; Vafai, K. A study of gravitational and magnetic effects on coupled stress bi-phase liquid suspended with crystal and Hafnium particles down in steep channel. *J. Mol. Liquids* **2019**, *286*, 110898. [CrossRef]
20. Prakash, J.; Tripathi, D.; Tiwari, A.K.; Sait, S.M.; Ellahi, R. Peristaltic pumping of nanofluids through a tapered channel in a porous environment: Applications in blood flow. *Symmetry* **2019**, *11*, 868. [CrossRef]
21. Riaz, A.; Bhatti, M.M.; Ellahi, R.; Zeeshan, A.; M Sait, S. Mathematical Analysis on an Asymmetrical Wavy Motion of Blood under the Influence Entropy Generation with Convective Boundary Conditions. *Symmetry* **2020**, *12*, 102. [CrossRef]
22. Gireesha, B.; Mahanthesh, B.; Gorla, R.S.R.; Manjunatha, P. Thermal radiation and Hall effects on boundary layer flow past a non-isothermal stretching surface embedded in porous medium with non-uniform heat source/sink and fluid-particle suspension. *Heat Mass Transf.* **2016**, *52*, 897–911. [CrossRef]
23. Bhatti, M.; Zeeshan, A.; Ijaz, N.; Bég, O.A.; Kadir, A. Mathematical modelling of nonlinear thermal radiation effects on EMHD peristaltic pumping of viscoelastic dusty fluid through a porous medium duct. *Eng. Sci. Technol. Int. J.* **2017**, *20*, 1129–1139. [CrossRef]
24. Kumar, K.G.; Rudraswamy, N.; Gireesha, B.; Manjunatha, S. Non linear thermal radiation effect on Williamson fluid with particle-liquid suspension past a stretching surface. *Results Phys.* **2017**, *7*, 3196–3202. [CrossRef]
25. Bhatti, M.; Zeeshan, A.; Ellahi, R.; Shit, G. Mathematical modeling of heat and mass transfer effects on MHD peristaltic propulsion of two-phase flow through a Darcy-Brinkman-Forchheimer porous medium. *Adv. Powder Technol.* **2018**, *29*, 1189–1197. [CrossRef]
26. Mahanthesh, B.; Gireesha, B. Scrutinization of thermal radiation, viscous dissipation and Joule heating effects on Marangoni convective two-phase flow of Casson fluid with fluid-particle suspension. *Results Phys.* **2018**, *8*, 869–878. [CrossRef]
27. Zhu, L.T.; Liu, Y.X.; Luo, Z.H. An enhanced correlation for gas-particle heat and mass transfer in packed and fluidized bed reactors. *Chem. Eng. J.* **2019**, *374*, 531–544. [CrossRef]

28. Bhatti, M.; Ellahi, R.; Zeeshan, A.; Marin, M.; Ijaz, N. Numerical study of heat transfer and Hall current impact on peristaltic propulsion of particle-fluid suspension with compliant wall properties. *Mod. Phys. Lett. B* **2019**, *33*, 1950439. [CrossRef]
29. Horne, A.W.; Critchley, H.O. Mechanisms of disease: The endocrinology of ectopic pregnancy. *Exp. Rev. Mol. Med.* **2012**, *14*. [CrossRef]
30. Srivastava, L.; Srivastava, V. Peristaltic transport of blood: Casson model—II. *J. Biomech.* **1984**, *17*, 821–829. [CrossRef]
31. Akbar, N.S.; Butt, A.W. Heat transfer analysis of Rabinowitsch fluid flow due to metachronal wave of cilia. *Results Phys.* **2015**, *5*, 92–98. [CrossRef]
32. Manzoor, N.; Bég, O.A.; Maqbool, K.; Shaheen, S. Mathematical modelling of ciliary propulsion of an electrically-conducting Johnson-Segalman physiological fluid in a channel with slip. *Comput. Methods Biomech. Biomed. Eng.* **2019**, 685-695. [CrossRef] [PubMed]
33. Chamkha, A.J.; Al-Subaie, M.A. Hydromagnetic buoyancy-induced flow of a particulate suspension through a vertical pipe with heat generation or absorption effects. *Turk. J. Eng. Environ. Sci.* **2010**, *33*, 127–134.
34. Marble, F.E. Dynamics of dusty gases. *Ann. Rev. Fluid Mech.* **1970**, *2*, 397–446. [CrossRef]

Article

Finite-Time Control for Nonlinear Systems with Time-Varying Delay and Exogenous Disturbance

Yanli Ruan [1,*] and Tianmin Huang [2]

[1] School of Electrical Engineering, Southwest Jiaotong University, Chengdu 610031, China
[2] School of Mathematics, Southwest Jiaotong University, Chengdu 610031, China; tmhuang@home.swjtu.edu.cn
* Correspondence: yanliruan666@163.com

Received: 31 December 2019; Accepted: 24 February 2020; Published: 11 March 2020

Abstract: This paper is concerned with the problem of finite-time control for nonlinear systems with time-varying delay and exogenous disturbance, which can be represented by a Takagi–Sugeno (T-S) fuzzy model. First, by constructing a novel augmented Lyapunov–Krasovskii functional involving several symmetric positive definite matrices, a new delay-dependent finite-time boundedness criterion is established for the considered T-S fuzzy time-delay system by employing an improved reciprocally convex combination inequality. Then, a memory state feedback controller is designed to guarantee the finite-time boundness of the closed-loop T-S fuzzy time-delay system, which is in the framework of linear matrix inequalities (LMIs). Finally, the effectiveness and merits of the proposed results are shown by a numerical example.

Keywords: finite-time boundedness; T-S fuzzy systems; time-varying delay; Lyapunov–Krasovskii functional (LKF)

---

1. Introduction

During the past several decades, the control problem of nonlinear systems has attracted considerable attention [1–6] as various practical systems are essentially nonlinear and cannot be easily simplified into a linear model. Up to now, many fuzzy logic control approaches have been proposed for the control problem of nonlinear systems. In particular, the Takagi–Sugeno (T-S) fuzzy model, developed in [7], is an important tool to approximate complex nonlinear systems by combining the fruitful linear system theory and the flexible fuzzy logic approach. Additionally, time-delay is unavoidably encountered in many dynamic systems, such as power systems, network control systems, neural networks, etc., which often results in chaos, oscillation, and even instability. Therefore, the study of T-S fuzzy time-delay systems has become more and more popular in recent years. In particular, many significant and interesting results on stability analysis and the control synthesis of T-S fuzzy time-delay systems have been developed in the literature [8–15].

Much attention has been paid to obtain the delay-dependent stability criteria for T-S fuzzy time-delay systems over the last few decades. It is well-known that the conservativeness of the stability criteria mainly has two sources: the choice of the Lyapunov–Krasovskii functional (LKF) and the estimation of its derivative. It is of great importance to construct an appropriate Lyapunov–Krasovskii functional for deriving less conservative stability conditions. In recent years, delay-partitioning Lyapunov–Krasovskii functionals and augmented Lyapunov–Krasovskii functionals have been developed to reduce the conservativeness of simple LKFs and have attracted growing attention. A delay-partitioning approach was applied to study the Lyapunov asymptotic stability of T-S fuzzy time-delay systems and some less conservative stability conditions were obtained in [16–18].

In [19], the authors introduced the triple-integral terms into the LKFs to derive the stability conditions for T-S fuzzy time-delay systems. In addition, various approaches have been proposed to estimate the derivatives of LKFs when dealing with stability analysis and control synthesis of time-delay systems, such as the free weighting matrix approach [20], Jensen inequality approach [21], Wirtinger-based integral inequality approach [10], reciprocally convex combination approach [22], auxiliary function-based inequality approach [23], and free-matrix-based integral inequality approach [24].

By applying the Wirtinger-based integral inequality approach and reciprocally convex combination approach, Zeng et al. [18] derived some less conservative stability criteria for uncertain T-S fuzzy systems with time-varying delays. An improved free weighting matrix approach was employed to obtain several new delay-dependent stability conditions in terms of the linear matrix inequalities for T-S fuzzy systems with time-varying delays in [25]. In [26], the authors investigated Lyapunov asymptotic stability analysis problems for T-S fuzzy time-delay systems by constructing a new augmented Lyapunov–Krasovskii functional and employing the free-matrix-based integral inequality approach.

The aforementioned results regarding the stability analysis of T-S fuzzy systems mainly focus on Lyapunov asymptotic stability, in which the states of systems converge asymptotically to equilibrium in an infinite time interval. However, in many practical engineering applications, the main concern may be the transient performances of the system trajectory during a specified finite-time interval. Unlike Lyapunov asymptotic stability, finite-time stability, introduced in [27], is another stability concept, which requires that the states of dynamical systems do not exceed a certain threshold in a fixed finite-time interval with a given bound for the initial condition. Up to now, the problem of finite-time stability, finite-time boundedness, and finite-time stabilization of dynamical systems has attracted growing attention, and many significant results have been reported in [28–32].

Several results on finite-time stability and stabilization of T-S fuzzy systems can also be found. The problem of finite-time stability and finite-time stabilization for T-S fuzzy time-delay systems was investigated in [28]. Sakthivel et al. [31] studied finite-time dissipative based fault-tolerant control problem for a class of T-S fuzzy systems with a constant delay. However, to the best of our knowledge, until now there have been few results on finite-time boundedness and finite-time stabilization of T-S fuzzy systems with time-varying delay and exogenous disturbance. Furthermore, it should be mentioned that most of the existing works on finite-time control for T-S fuzzy time-delay systems are fairly conservative. Motivated by the above discussions, in this paper, we deal with the problem of finite-time control for a class of nonlinear systems with time-varying delay and exogenous disturbance, which can be described by a T-S fuzzy model.

The main contributions of this paper are summarized as follows. First, a new augmented Lyapunov–Krasovskii functional is constructed, which makes full use of the information about time-varying delay. Based on the proposed Lyapunov–Krasovskii functional, a less conservative finite-time boundedness condition is obtained for T-S fuzzy time-delay systems by utilizing an improved reciprocally convex combination technique. Second, based on parallel distributed compensation schemes, a memory state feedback controller is designed to finite-time stabilize the T-S fuzzy time-delay system, which can be derived by solving a series of linear matrix inequalities (LMIs). Finally, a numerical example is given to illustrate the advantages and validity of the developed results.

The rest of this paper is organized as follows: the problem statement is given in Section 2. The main results on the finite-time boundedness and finite-time stabilization of nonlinear systems with time-varying delay and exogenous disturbance are presented in Section 3. In Section 4, a numerical example is proposed to show the effectiveness of the developed results. Finally, our conclusions are drawn in Section 5.

Notations: Throughout this paper, $\mathbf{R}^n$ denotes the n-dimensional Euclidean space; $\mathbf{R}^{n\times m}$ stands for the set of all $n \times m$ real matrices; the superscripts T and -1 denote the transpose and inverse of a matrix, respectively; I and 0 represent the identity matrix and zero matrix, respectively, with compatible dimensions; $diag\{\cdots\}$ denotes a block-diagonal matrix; the notation $P > 0 (\geq 0)$ means that the matrix

P is real symmetric and positive definite (semi-positive definite); $*$ stands for the symmetric terms in a symmetric matrix; for any matrix $X \in \mathbf{R}^{n\times n}$, $Sym\{X\}$ is defined as $X + X^T$.

2. Problem Formulation

Consider a class of nonlinear systems with time-varying delay and exogenous disturbance, which can be represented by the following T-S fuzzy model:

Plant Rule i:

IF $\xi_1(t)$ is $N_{i1}, \cdots$, and $\xi_p(t)$ is N_{ip},

THEN

$$\begin{cases} \dot{x}(t) = A_i x(t) + A_{di} x(t - d(t)) + B_i u(t) + G_i \omega(t) \\ x(t) = \phi(t), \quad t \in [-h, 0] \end{cases}$$

where $i \in \{1, 2, \ldots, r\}$, r is the number of IF-THEN rules, $x(t) \in \mathbf{R}^n$ is the state vector, $u(t) \in \mathbf{R}^p$ is the control input, $\omega(t) \in \mathbf{R}^l$ is the exogenous disturbance, which satisfies $\int_0^{T_f} \omega(t)^T \omega(t) dt \le \delta$; $\delta \ge 0$ is a given scalar. A_i, A_{di}, B_i, and G_i are known constant matrices with appropriate dimensions. $\xi_1(t), \xi_2(t), \ldots, \xi_p(t)$ are premise variables, $N_{i1}, N_{i2}, \ldots, N_{ip}$ are fuzzy sets. The time delay $d(t)$ is a time-varying function that satisfies

$$0 \le d(t) \le h \quad and \quad \mu_1 \le \dot{d}(t) \le \mu_2 \tag{1}$$

where $h > 0$ and μ_1, μ_2 are constants. The initial condition $\phi(t)$ is a continuous vector-valued function for all $t \in [-h, 0]$.

Let $\xi(t) = [\xi_1(t), \xi_2(t), \ldots, \xi_p(t)]^T$, by employing a singleton fuzzifier, product inference, and center-average defuzzifer, the input–output form of the above T-S fuzzy time-delay system can be represented by

$$\dot{x}(t) = \sum_{i=1}^{r} \rho_i(\xi(t))\{A_i x(t) + A_{di} x(t - d(t)) + B_i u(t) + G_i \omega(t)\} \tag{2}$$

where $\rho_i(\xi(t)) = \frac{\theta_i(\xi(t))}{\sum_{i=1}^{r} \theta_i(\xi(t))}$, $\theta_i(\xi(t)) = \prod_{j=1}^{p} N_{ij}(\xi_j(t))$, $N_{ij}(\xi_j(t))$ is the grade of membership of $\xi_j(t)$ in the fuzzy set N_{ij}. We note that $\theta_i(\xi(t)) \ge 0$, $\sum_{i=1}^{r} \theta_i(\xi(t)) > 0$ for all t, and we can obtain $\rho_i(\xi(t)) \ge 0$, $\sum_{i=1}^{r} \rho_i(\xi(t)) = 1$.

In this paper, for simplicity, we denote $S(t) = \sum_{i=1}^{r} \rho_i(\xi(t)) S_i$ for any matrix S_i. Therefore, the T-S fuzzy time-delay system (2) can be rewritten as follows:

$$\dot{x}(t) = A(t)x(t) + A_d(t)x(t - d(t)) + B(t)u(t) + G(t)\omega(t). \tag{3}$$

Now, the definition of finite-time boundedness (FTB) for the T-S fuzzy time-delay system (3) with $u(t) \equiv 0$ is given as follows:

Definition 1 ([31]). *The T-S fuzzy time-delay system (3) with $u(t) \equiv 0$ is said to be finite-time bounded with respect to $(c_1, c_2, T_f, R, \delta, h)$, where $0 < c_1 < c_2$, $T_f > 0$, $R \in \mathbf{R}^{n\times n}$ and $R > 0$, if*

$$sup_{-h \le \theta \le 0}\{x^T(\theta)Rx(\theta), \dot{x}^T(\theta)R\dot{x}(\theta)\} \le c_1 \Rightarrow x^T(t)Rx(t) < c_2,$$

$\forall t \in [0, T_f], \forall \omega(t) : \int_0^{T_f} \omega(t)^T \omega(t) dt \le \delta$.

Based on the parallel distributed compensation scheme, we aim to design the following memory state feedback controller, which can guarantee the corresponding closed-loop T-S fuzzy time-delay system finite-time bounded:

$$u(t) = K_1(t)x(t) + K_2(t)x(t-d(t)), \tag{4}$$

where $K_1(t) = \sum_{j=1}^{r} \rho_j(\xi(t))K_{1j}$, $K_2(t) = \sum_{j=1}^{r} \rho_j(\xi(t))K_{2j}$, and $K_{1j}, K_{2j}, j = 1, 2, \ldots, r$ are the state feedback gain matrices to be determined.

By substituting (4) into (3), the corresponding closed-loop T-S fuzzy time-delay system can be represented as follows:

$$\dot{x}(t) = [A(t) + B(t)K_1(t)]x(t) + [A_d(t) + B(t)K_2(t)]x(t-d(t)) + G(t)\omega(t). \tag{5}$$

In order to derive the main results in this paper, the following lemma, i.e., the improved reciprocally convex combination inequality approach will be utilized in finite-time boundedness analysis and controller design of T-S fuzzy time-delay systems.

Lemma 1 ([33]). *Let $R_1, R_2 \in \mathbf{R}^{m\times m}$ be real symmetric positive definite matrices, $\varpi_1, \varpi_2 \in \mathbf{R}^m$ and a scalar $\alpha \in (0,1)$. Then for any matrices $Y_1, Y_2 \in \mathbf{R}^{m\times m}$, the following inequality holds*

$$\begin{aligned}
&\frac{1}{\alpha}\varpi_1^T R_1 \varpi_1 + \frac{1}{1-\alpha}\varpi_2^T R_2 \varpi_2 \\
\geq &\varpi_1^T[R_1 + (1-\alpha)(R_1 - Y_1 R_2^{-1} Y_1^T)]\varpi_1 \\
&+ \varpi_2^T[R_2 + \alpha(R_2 - Y_2^T R_1^{-1} Y_2)]\varpi_2 \\
&+ 2\varpi_1^T[\alpha Y_1 + (1-\alpha)Y_2]\varpi_2.
\end{aligned}$$

3. Main Results

3.1. Finite-Time Boundedness Analysis

In this subsection, our aim is to develop a new delay-dependent finite-time boundedness criterion for T-S fuzzy systems with time-varying delay and norm-bounded disturbance. Before deriving the main results, the nomenclature simplifying the representations for matrices and vectors is given as follows:

$$\varepsilon_1(t) = \begin{pmatrix} x(t) \\ x(t-d(t)) \\ x(t-h) \\ \dot{x}(t-d(t)) \\ \dot{x}(t-h) \end{pmatrix}, \quad \varepsilon_2(t) = \begin{pmatrix} \frac{1}{d(t)}\int_{t-d(t)}^{t} x(s)ds \\ \frac{1}{h-d(t)}\int_{t-h}^{t-d(t)} x(s)ds \\ \frac{1}{d^2(t)}\int_{t-d(t)}^{t}\int_{\theta}^{t} x(s)dsd\theta \\ \frac{1}{(h-d(t))^2}\int_{t-h}^{t-d(t)}\int_{\theta}^{t-d(t)} x(s)dsd\theta \\ \frac{1}{h}\int_{t-h}^{t} x(s)ds \\ \frac{1}{h^2}\int_{t-h}^{t}\int_{\theta}^{t} x(s)dsd\theta \\ \omega(t) \end{pmatrix},$$

$$\varepsilon(t) = \begin{bmatrix} \varepsilon_1^T(t) & \varepsilon_2^T(t) \end{bmatrix}^T, \quad \bar{\varepsilon}(t) = \begin{bmatrix} \varepsilon_1^T(t) & \varepsilon_2^T(t) & \dot{x}^T(t) \end{bmatrix}^T,$$

$$e_i = \begin{bmatrix} 0_{n\times(i-1)n} & I_n & 0_{n\times(12-i)n} \end{bmatrix}, \quad i = 1, 2, \cdots, 12,$$

$$\tilde{e}_i = \begin{bmatrix} 0_{n\times(i-1)n} & I_n & 0_{n\times(13-i)n} \end{bmatrix}, \quad i = 1,2,\cdots,13.$$

Theorem 1. *For given scalars $h > 0$ and μ_1, μ_2, the T-S fuzzy system (3) with $u(t) = 0$ and a time-varying delay $d(t)$ satisfying (1) is finite-time bounded with respect to $(c_1, c_2, T_f, R, \delta, h)$, if there exists a scalar $\beta > 0$, symmetric positive definite matrices $P \in \mathbf{R}^{5n\times 5n}$, $S_1, S_2 \in \mathbf{R}^{2n\times 2n}$, $Q_1, Q_2 \in \mathbf{R}^{3n\times 3n}$, $W, Z, U \in \mathbf{R}^{n\times n}$, and any matrices $Y_1, Y_2 \in \mathbf{R}^{3n\times 3n}$, such that the following conditions hold:*

$$\begin{pmatrix} \Sigma_{1i}(0,\mu_1) & \Lambda_1^T Y_1 \\ Y_1^T \Lambda_1 & -W_0 \end{pmatrix} < 0, \quad i = 1,2,\ldots,r \tag{6}$$

$$\begin{pmatrix} \Sigma_{1i}(0,\mu_2) & \Lambda_1^T Y_1 \\ Y_1^T \Lambda_1 & -W_0 \end{pmatrix} < 0, \quad i = 1,2,\ldots,r \tag{7}$$

$$\begin{pmatrix} \Sigma_{1i}(h,\mu_1) & \Lambda_2^T Y_2^T \\ Y_2 \Lambda_2 & -W_0 \end{pmatrix} < 0, \quad i = 1,2,\ldots,r \tag{8}$$

$$\begin{pmatrix} \Sigma_{1i}(h,\mu_2) & \Lambda_2^T Y_2^T \\ Y_2 \Lambda_2 & -W_0 \end{pmatrix} < 0, \quad i = 1,2,\ldots,r \tag{9}$$

$$c_1\Pi + \lambda_{37}\delta < \lambda_{36} c_2 e^{-\beta T_f}, \tag{10}$$

where

$$\begin{aligned} P &= \begin{pmatrix} P_{11} & P_{12} & P_{13} & P_{14} & P_{15} \\ * & P_{22} & P_{23} & P_{24} & P_{25} \\ * & * & P_{33} & P_{34} & P_{35} \\ * & * & * & P_{44} & P_{45} \\ * & * & * & * & P_{55} \end{pmatrix}, Q_1 = \begin{pmatrix} Q_{11} & Q_{12} & Q_{13} \\ * & Q_{22} & Q_{23} \\ * & * & Q_{33} \end{pmatrix}, Q_2 = \begin{pmatrix} q_{11} & q_{12} & q_{13} \\ * & q_{22} & q_{23} \\ * & * & q_{33} \end{pmatrix}, \\ S_1 &= \begin{pmatrix} S_{11} & S_{12} \\ * & S_{22} \end{pmatrix}, S_2 = \begin{pmatrix} s_{11} & s_{12} \\ * & s_{22} \end{pmatrix}, \end{aligned}$$

$$\begin{aligned} \Sigma_{1i}(d(t),\dot{d}(t)) = & Sym\{\Xi_1^T P \Xi_{2i}\} + \dot{d}(t)\Xi_3^T S_1 \Xi_3 - \dot{d}(t)\Xi_4^T S_2 \Xi_4 + Sym(\Xi_3^T S_1 \Xi_{5i} + \Xi_4^T S_2 \Xi_{6i}) \\ & + Sym(\Xi_7^T Q_1 \Xi_{8i}) + \Xi_{9i}^T Q_1 \Xi_{9i} - (1-\dot{d}(t))\Xi_{10}^T Q_1 \Xi_{10} + Sym(\Xi_{11}^T Q_2 \Xi_{12}) \\ & + (1-\dot{d}(t))\Xi_{13}^T Q_2 \Xi_{13} - \Xi_{14}^T Q_2 \Xi_{14} + h^2 e_{si}^T W e_{si} + \frac{h^4}{4} e_{si}^T Z e_{si} - h^2 \Xi_{15}^T Z \Xi_{15} \\ & - 2h^2 \Xi_{16}^T Z \Xi_{16} - e_{12}^T U e_{12} + (\alpha-2)\Lambda_1^T W_0 \Lambda_1 - (\alpha+1)\Lambda_2^T W_0 \Lambda_2 \\ & - Sym\{\Lambda_1^T [\alpha Y_1 + (1-\alpha) Y_2] \Lambda_2\}, \end{aligned}$$

$$\begin{aligned} \alpha &= \frac{d(t)}{h}, \qquad W_0 = diag\{W, 3W, 5W\}, \qquad \Xi_1 = \begin{bmatrix} e_1^T & e_2^T & e_3^T & d(t)e_6^T & (h-d(t))e_7^T \end{bmatrix}^T, \\ \Xi_{2i} &= [e_{si}^T \quad (1-\dot{d}(t))e_4^T \quad e_5^T \quad e_1^T - (1-\dot{d}(t))e_2^T \quad (1-\dot{d}(t))e_2^T - e_3^T]^T, \\ \Xi_3 &= \begin{bmatrix} e_1^T & e_6^T \end{bmatrix}^T, \qquad \Xi_4 = \begin{bmatrix} e_1^T & e_7^T \end{bmatrix}^T, \qquad \Xi_{5i} = \begin{bmatrix} d(t)e_{si}^T & -\dot{d}(t)e_6^T + e_1^T - (1-\dot{d}(t))e_2^T \end{bmatrix}^T, \end{aligned}$$

$$
\begin{aligned}
\Xi_{6i} &= \left[(h-d(t))e_{si}^T \quad \dot{d}(t)e_7^T + (1-\dot{d}(t))e_2^T - e_3^T\right]^T, \ \Xi_7 = \left[d(t)e_6^T \quad e_1^T - e_2^T \quad d(t)(e_1^T - e_6^T)\right]^T, \\
\Xi_{8i} &= \left[0 \quad 0 \quad e_{si}^T\right]^T, \ \Xi_{9i} = \left[e_1^T \quad e_{si}^T \quad 0\right]^T, \ \Xi_{10} = \left[e_2^T \quad e_4^T \quad e_1^T - e_2^T\right]^T, \\
\Xi_{11} &= [(h-d(t))e_7^T \quad e_2^T - e_3^T \quad (h-d(t))(e_2^T - e_7^T)]^T, \ \Xi_{12} = \left[0 \quad 0 \quad (1-\dot{d}(t))e_4^T\right]^T, \\
\Xi_{13} &= \left[e_2^T \quad e_4^T \quad 0\right]^T, \ \Xi_{14} = \left[e_3^T \quad e_5^T \quad e_2^T - e_3^T\right]^T, \ \Xi_{15} = e_1 - e_{10}, \ \Xi_{16} = e_1 + 2e_{10} - 6e_{11}, \\
\Lambda_1 &= [e_1^T - e_2^T \quad e_1^T + e_2^T - 2e_6^T \quad e_1^T - e_2^T + 6e_6^T - 12e_8^T]^T, \\
\Lambda_2 &= [e_2^T - e_3^T \quad e_2^T + e_3^T - 2e_7^T \quad e_2^T - e_3^T + 6e_7^T - 12e_9^T]^T, \\
e_{si} &= A_i e_1 + A_{di} e_2 + G_i e_{12}, \\
\Pi &= \lambda_1 + \lambda_2 + \lambda_3 + h^2(\lambda_4 + \lambda_5 + \lambda_{26} + \lambda_{27} + \lambda_{32} + \lambda_{33}) + 2(\lambda_6 + \lambda_7 + \lambda_{10}) \\
&\quad +2h(\lambda_8 + \lambda_9 + \lambda_{11} + \lambda_{12} + \lambda_{13} + \lambda_{14} + \lambda_{18} + \lambda_{21} + \lambda_{25} + \lambda_{31}) + 2h^2\lambda_{15} \\
&\quad +h(\lambda_{16} + \lambda_{17} + \lambda_{19} + \lambda_{20} + \lambda_{22} + \lambda_{23} + \lambda_{28} + \lambda_{29}) + \frac{h^3}{3}(\lambda_{24} + \lambda_{30}) \\
&\quad +\frac{h^3}{2}\lambda_{34} + \frac{h^5}{12}\lambda_{35}, \\
\lambda_1 &= \lambda_{max}(\bar{P}_{11}), \ \lambda_2 = \lambda_{max}(\bar{P}_{22}), \ \lambda_3 = \lambda_{max}(\bar{P}_{33}), \ \lambda_4 = \lambda_{max}(\bar{P}_{44}), \ \lambda_5 = \lambda_{max}(\bar{P}_{55}), \\
\lambda_6 &= \lambda_{max}(\bar{P}_{12}), \ \lambda_7 = \lambda_{max}(\bar{P}_{13}), \ \lambda_8 = \lambda_{max}(\bar{P}_{14}), \ \lambda_9 = \lambda_{max}(\bar{P}_{15}), \ \lambda_{10} = \lambda_{max}(\bar{P}_{23}), \\
\lambda_{11} &= \lambda_{max}(\bar{P}_{24}), \ \lambda_{12} = \lambda_{max}(\bar{P}_{25}), \ \lambda_{13} = \lambda_{max}(\bar{P}_{34}), \ \lambda_{14} = \lambda_{max}(\bar{P}_{35}), \ \lambda_{15} = \lambda_{max}(\bar{P}_{45}), \\
\lambda_{16} &= \lambda_{max}(\bar{S}_{11}), \ \lambda_{17} = \lambda_{max}(\bar{S}_{22}), \ \lambda_{18} = \lambda_{max}(\bar{S}_{12}), \ \lambda_{19} = \lambda_{max}(\bar{s}_{11}), \ \lambda_{20} = \lambda_{max}(\bar{s}_{22}), \\
\lambda_{21} &= \lambda_{max}(\bar{s}_{12}), \ \lambda_{22} = \lambda_{max}(\bar{Q}_{11}), \ \lambda_{23} = \lambda_{max}(\bar{Q}_{22}), \ \lambda_{24} = \lambda_{max}(\bar{Q}_{33}), \ \lambda_{25} = \lambda_{max}(\bar{Q}_{12}), \\
\lambda_{26} &= \lambda_{max}(\bar{Q}_{13}), \ \lambda_{27} = \lambda_{max}(\bar{Q}_{23}), \ \lambda_{28} = \lambda_{max}(\bar{q}_{11}), \ \lambda_{29} = \lambda_{max}(\bar{q}_{22}), \ \lambda_{30} = \lambda_{max}(\bar{q}_{33}), \\
\lambda_{31} &= \lambda_{max}(\bar{q}_{12}), \ \lambda_{32} = \lambda_{max}(\bar{q}_{13}), \ \lambda_{33} = \lambda_{max}(\bar{q}_{23}), \ \lambda_{34} = \lambda_{max}(\bar{W}), \ \lambda_{35} = \lambda_{max}(\bar{Z}), \\
\lambda_{36} &= \lambda_{min}(\bar{P}_{11}), \ \lambda_{37} = \lambda_{max}(U), \\
\bar{P}_{1j} &= R^{-\frac{1}{2}}P_{1j}R^{-\frac{1}{2}}, \ j = 1,2,3,4,5, \ \bar{P}_{2j} = R^{-\frac{1}{2}}P_{2j}R^{-\frac{1}{2}}, \ j = 2,3,4,5, \\
\bar{P}_{3j} &= R^{-\frac{1}{2}}P_{3j}R^{-\frac{1}{2}}, \ j = 3,4,5, \ \bar{P}_{4j} = R^{-\frac{1}{2}}P_{4j}R^{-\frac{1}{2}}, \ j = 4,5, \ \bar{P}_{55} = R^{-\frac{1}{2}}P_{55}R^{-\frac{1}{2}}, \\
\bar{S}_{1j} &= R^{-\frac{1}{2}}S_{1j}R^{-\frac{1}{2}}, \ j = 1,2, \ \bar{S}_{22} = R^{-\frac{1}{2}}S_{22}R^{-\frac{1}{2}}, \ \bar{s}_{1j} = R^{-\frac{1}{2}}s_{1j}R^{-\frac{1}{2}}, \ j = 1,2, \ \bar{s}_{22} = R^{-\frac{1}{2}}s_{22}R^{-\frac{1}{2}}, \\
\bar{Q}_{1j} &= R^{-\frac{1}{2}}Q_{1j}R^{-\frac{1}{2}}, \ j = 1,2,3, \ \bar{Q}_{2j} = R^{-\frac{1}{2}}Q_{2j}R^{-\frac{1}{2}}, \ j = 2,3, \ \bar{Q}_{33} = R^{-\frac{1}{2}}Q_{33}R^{-\frac{1}{2}}, \\
\bar{q}_{1j} &= R^{-\frac{1}{2}}q_{1j}R^{-\frac{1}{2}}, \ j = 1,2,3, \ \bar{q}_{2j} = R^{-\frac{1}{2}}q_{2j}R^{-\frac{1}{2}}, \ j = 2,3, \ \bar{q}_{33} = R^{-\frac{1}{2}}q_{33}R^{-\frac{1}{2}}, \\
\bar{W} &= R^{-\frac{1}{2}}WR^{-\frac{1}{2}}, \ \bar{Z} = R^{-\frac{1}{2}}ZR^{-\frac{1}{2}}.
\end{aligned}
$$

Proof. We construct the following Lyapunov–Krasovskii functional candidate for the T-S fuzzy time-delay system (3):

$$V(x(t)) = V_1(x(t)) + V_2(x(t)) + V_3(x(t)) + V_4(x(t)) + V_5(x(t)) + V_6(x(t)), \tag{11}$$

where

$$V_1(x(t)) = \eta_1^T(t)P\eta_1(t),$$

$$V_2(x(t)) = d(t)\eta_2^T(t)S_1\eta_2(t) + (h-d(t))\eta_3^T(t)S_2\eta_3(t),$$

$$V_3(x(t)) = \int_{t-d(t)}^{t} \eta_4^T(s)Q_1\eta_4(s)ds,$$

$$V_4(x(t)) = \int_{t-h}^{t-d(t)} \eta_5^T(s)Q_2\eta_5(s)ds,$$

$V_5(x(t)) = h\int_{t-h}^{t}\int_{\theta}^{t}\dot{x}^T(s)W\dot{x}(s)dsd\theta,$

$V_6(x(t)) = \frac{h^2}{2}\int_{t-h}^{t}\int_{\sigma}^{t}\int_{\theta}^{t}\dot{x}^T(s)Z\dot{x}(s)dsd\theta d\sigma,$

and

$\eta_1(t) = [x^T(t) \quad x^T(t-d(t)) \quad x^T(t-h) \quad \int_{t-d(t)}^{t} x^T(s)ds \quad \int_{t-h}^{t-d(t)} x^T(s)ds]^T,$

$\eta_2(t) = [x^T(t) \quad \frac{1}{d(t)}\int_{t-d(t)}^{t} x^T(s)ds]^T,$

$\eta_3(t) = [x^T(t) \quad \frac{1}{h-d(t)}\int_{t-h}^{t-d(t)} x^T(s)ds]^T,$

$\eta_4(s) = [x^T(s) \quad \dot{x}^T(s) \quad \int_{s}^{t}\dot{x}^T(\theta)d\theta]^T,$

$\eta_5(s) = [x^T(s) \quad \dot{x}^T(s) \quad \int_{s}^{t-d(t)}\dot{x}^T(\theta)d\theta]^T.$

Then, the time derivatives of $V_i(x(t))(i = 1,2,3,4,5,6)$ along the trajectory of the T-S fuzzy system (3) are obtained as follows:

$$\begin{aligned}\dot{V}_1(x(t)) &= 2\begin{pmatrix} x(t) \\ x(t-d(t)) \\ x(t-h) \\ \int_{t-d(t)}^{t} x(s)ds \\ \int_{t-h}^{t-d(t)} x(s)ds \end{pmatrix}^T P \begin{pmatrix} \dot{x}(t) \\ (1-\dot{d}(t))\dot{x}(t-d(t)) \\ \dot{x}(t-h) \\ x(t)-(1-\dot{d}(t))x(t-d(t)) \\ (1-\dot{d}(t))x(t-d(t))-x(t-h) \end{pmatrix} \\ &= \sum_{i=1}^{r}\rho_i(\xi(t))\varepsilon^T(t)[Sym(\Xi_1^T P\Xi_{2i})]\varepsilon(t).\end{aligned} \tag{12}$$

Similarly, we can also obtain

$$\dot{V}_2(x(t)) = \sum_{i=1}^{r}\rho_i(\xi(t))\varepsilon^T(t)[\dot{d}(t)\Xi_3^T S_1\Xi_3 - \dot{d}(t)\Xi_4^T S_2\Xi_4 + Sym(\Xi_3^T S_1\Xi_{5i} + \Xi_4^T S_2\Xi_{6i})]\varepsilon(t), \tag{13}$$

$$\dot{V}_3(x(t)) = \sum_{i=1}^{r}\rho_i(\xi(t))\varepsilon^T(t)[Sym(\Xi_7^T Q_1\Xi_{8i}) + \Xi_{9i}^T Q_1\Xi_{9i} - (1-\dot{d}(t))\Xi_{10}^T Q_1\Xi_{10}]\varepsilon(t), \tag{14}$$

$$\dot{V}_4(x(t)) = \sum_{i=1}^{r}\rho_i(\xi(t))\varepsilon^T(t)[Sym(\Xi_{11}^T Q_2\Xi_{12}) + (1-\dot{d}(t))\Xi_{13}^T Q_2\Xi_{13} - \Xi_{14}^T Q_2\Xi_{14}]\varepsilon(t), \tag{15}$$

$$\begin{aligned}\dot{V}_5(x(t)) &= h^2\dot{x}^T(t)W\dot{x}(t) - h\int_{t-h}^{t}\dot{x}^T(s)W\dot{x}(s)ds \\ &= \sum_{i=1}^{r}\rho_i(\xi(t))\varepsilon^T(t)(h^2 e_{si}^T W e_{si})\varepsilon(t) \\ &\quad - h\int_{t-h}^{t}\dot{x}^T(s)W\dot{x}(s)ds,\end{aligned} \tag{16}$$

$$\begin{aligned}\dot{V}_6(x(t)) &= \frac{h^4}{4}\dot{x}^T(t)Z\dot{x}(t) - \frac{h^2}{2}\int_{t-h}^{t}\int_{\theta}^{t}\dot{x}^T(s)Z\dot{x}(s)dsd\theta \\ &= \sum_{i=1}^{r}\rho_i(\xi(t))\varepsilon^T(t)(\frac{h^4}{4}e_{si}^T Z e_{si})\varepsilon(t) \\ &\quad - \frac{h^2}{2}\int_{t-h}^{t}\int_{\theta}^{t}\dot{x}^T(s)Z\dot{x}(s)dsd\theta.\end{aligned} \tag{17}$$

Now, we split $-h\int_{t-h}^{t}\dot{x}^T(s)W\dot{x}(s)ds$ into two integrals, i.e., $-h\int_{t-h}^{t}\dot{x}^T(s)W\dot{x}(s)ds =-h\int_{t-d(t)}^{t}\dot{x}^T(s)W\dot{x}(s)ds$

$-h\int_{t-h}^{t-d(t)}\dot{x}^T(s)W\dot{x}(s)ds$. Then, utilizing the integral inequality (24) in Lemma 5.1 of [23] for each of them yields

$$-h\int_{t-d(t)}^{t}\dot{x}^T(s)W\dot{x}(s)ds \le -\frac{h}{d(t)}\varepsilon^T(t)\Lambda_1^T W_0\Lambda_1\varepsilon(t) \tag{18}$$

and

$$-h\int_{t-h}^{t-d(t)}\dot{x}^T(s)W\dot{x}(s)ds \le -\frac{h}{h-d(t)}\varepsilon^T(t)\Lambda_2^T W_0\Lambda_2\varepsilon(t), \tag{19}$$

where $W_0 = diag\{W, 3W, 5W\}$, $\Lambda_1 = \begin{pmatrix} e_1 - e_2 \\ e_1 + e_2 - 2e_6 \\ e_1 - e_2 + 6e_6 - 12e_8 \end{pmatrix}$ and $\Lambda_2 = \begin{pmatrix} e_2 - e_3 \\ e_2 + e_3 - 2e_7 \\ e_2 - e_3 + 6e_7 - 12e_9 \end{pmatrix}$.

According to Lemma 1, let $\alpha = \frac{d(t)}{h}$, $R_1 = R_2 = W_0$, $\varpi_1 = \Lambda_1\varepsilon(t)$, $\varpi_2 = \Lambda_2\varepsilon(t)$, from inequalities (18) and (19), then we can obtain

$$\begin{aligned} &-h\int_{t-d(t)}^{t}\dot{x}^T(s)W\dot{x}(s)ds - h\int_{t-h}^{t-d(t)}\dot{x}^T(s)W\dot{x}(s)ds \\ \le&\varepsilon^T(t)[(\alpha-2)\Lambda_1^T W_0\Lambda_1 - (\alpha+1)\Lambda_2^T W_0\Lambda_2 - Sym\{\Lambda_1^T[\alpha Y_1 + (1-\alpha)Y_2]\Lambda_2\} \\ &+ (1-\alpha)\Lambda_1^T Y_1 W_0^{-1} Y_1^T\Lambda_1 + \alpha\Lambda_2^T Y_2^T W_0^{-1} Y_2\Lambda_2]\varepsilon(t). \end{aligned} \tag{20}$$

Applying the integral inequality (25) in Lemma 5.1 of [23] to the double integral $-\frac{h^2}{2}\int_{t-h}^{t}\int_{\theta}^{t}\dot{x}^T(s)Z\dot{x}(s)dsd\theta$ in inequality (17) leads to

$$-\frac{h^2}{2}\int_{t-h}^{t}\int_{\theta}^{t}\dot{x}^T(s)Z\dot{x}(s)dsd\theta \le \varepsilon(t)^T(-h^2\Xi_{15}^T Z\Xi_{15} - 2h^2\Xi_{16}^T Z\Xi_{16})\varepsilon(t), \tag{21}$$

where $\Xi_{15} = e_1 - e_{10}$, $\Xi_{16} = e_1 + 2e_{10} - 6e_{11}$.

Notice that $\sum_{i=1}^{r}\rho_i(\xi(t)) = 1$, and we can derive the following result from (12)–(17), (20) and (21):

$$\dot{V}(x(t)) \le \sum_{i=1}^{r}\rho_i(\xi(t))\varepsilon^T(t)\Sigma_i(d(t),\dot{d}(t))\varepsilon(t) + \omega^T(t)U\omega(t), \tag{22}$$

where $\Sigma_i(d(t),\dot{d}(t)) = \Sigma_{1i}(d(t),\dot{d}(t)) + \Sigma_2(d(t))$,

$$\begin{aligned} \Sigma_{1i}(d(t),\dot{d}(t)) =& Sym\{\Xi_1^T P\Xi_{2i}\} + \dot{d}(t)\Xi_3^T S_1\Xi_3 - \dot{d}(t)\Xi_4^T S_2\Xi_4 + Sym(\Xi_3^T S_1\Xi_{5i} + \Xi_4^T S_2\Xi_{6i}) \\ &+ Sym(\Xi_7^T Q_1\Xi_{8i}) + \Xi_{9i}^T Q_1\Xi_{9i} - (1-\dot{d}(t))\Xi_{10}^T Q_1\Xi_{10} + Sym(\Xi_{11}^T Q_2\Xi_{12}) \\ &+ (1-\dot{d}(t))\Xi_{13}^T Q_2\Xi_{13} - \Xi_{14}^T Q_2\Xi_{14} + h^2 e_{si}^T We_{si} + \frac{h^4}{4}e_{si}^T Ze_{si} - h^2\Xi_{15}^T Z\Xi_{15} \\ &- 2h^2\Xi_{16}^T Z\Xi_{16} - e_{12}^T Ue_{12} + (\alpha-2)\Lambda_1^T W_0\Lambda_1 - (\alpha+1)\Lambda_2^T W_0\Lambda_2 \\ &- Sym\{\Lambda_1^T[\alpha Y_1 + (1-\alpha)Y_2]\Lambda_2\}, \end{aligned}$$

$$\Sigma_2(d(t)) = (1-\alpha)\Lambda_1^T Y_1 W_0^{-1} Y_1^T\Lambda_1 + \alpha\Lambda_2^T Y_2^T W_0^{-1} Y_2\Lambda_2.$$

Assuming $\Sigma_i(d(t),\dot{d}(t)) < 0$ for $i = 1, 2, \cdots, r$, we have

$$\dot{V}(x(t)) < \beta V(x(t)) + \omega^T(t)U\omega(t), \tag{23}$$

where $\beta > 0$ is a constant.

However, $\Sigma_i(d(t),\dot{d}(t))$ depends on the time-varying delay $d(t)$ and its derivative $\dot{d}(t)$. Therefore, $\Sigma_i(d(t),\dot{d}(t)) < 0$ cannot be solved directly by applying an LMI tool. Noting that $\Sigma_i(d(t),\dot{d}(t))$ is a linear function of $d(t)$ and $\dot{d}(t)$, it is obvious that $\Sigma_i(d(t),\dot{d}(t)) < 0$ can be satisfied if the following inequalities (24)–(27) hold,

$$\Sigma_i(0,\mu_1) < 0, \tag{24}$$

$$\Sigma_i(0,\mu_2) < 0, \tag{25}$$

$$\Sigma_i(h,\mu_1) < 0, \tag{26}$$

$$\Sigma_i(h,\mu_2) < 0. \tag{27}$$

According to Schur complement lemma, the inequalities (24)–(27) are equivalent to inequalities (6)–(9), respectively. Thus, the inequalities (6)–(9) can ensure $\Sigma_i(d(t),\dot{d}(t)) < 0$ holds. Furthermore, the inequalities (6)–(9) can also guarantee that the inequality (23) holds.

Multiplying (23) by $e^{-\beta t}$, we can obtain

$$e^{-\beta t}\dot{V}(x(t)) - \beta e^{-\beta t}V(x(t)) < e^{-\beta t}\omega^T(t)U\omega(t),$$

i.e.,

$$\frac{d}{dt}(e^{-\beta t}V(x(t))) < e^{-\beta t}\omega^T(t)U\omega(t). \tag{28}$$

Integrating (28) from 0 to t with $t \in [0, T_f]$, we have

$$e^{-\beta t}V(x(t)) - V(x(0)) < \int_0^t e^{-\beta s}\omega^T(s)U\omega(s)ds.$$

Noting that $\beta > 0$, we can derive

$$\begin{aligned} V(x(t)) &< e^{\beta t}V(x(0)) + e^{\beta t}\int_0^t e^{-\beta s}\omega^T(s)U\omega(s)ds \\ &\leq e^{\beta t}V(x(0)) + e^{\beta t}\lambda_{max}(U)\int_0^t \omega^T(s)\omega(s)ds. \end{aligned}$$

Therefore, we have

$$V(x(t)) < e^{\beta T_f}\left[V(x(0)) + \lambda_{max}(U)\delta\right]. \tag{29}$$

In addition, it can be easily obtained that

$$V(x(t)) \geq x^T(t)P_{11}x(t) = x^T(t)R^{\frac{1}{2}}\bar{P}_{11}R^{\frac{1}{2}}x(t) \geq \lambda_{min}(\bar{P}_{11})x^T(t)Rx(t) = \lambda_{36}x^T(t)Rx(t),$$

$$\begin{aligned} V(x(0)) &= \eta_1^T(0)P\eta_1(0) + d(0)\eta_2^T(0)S_1\eta_2(0) + (h-d(0))\eta_3^T(0)S_2\eta_3(0) + \int_{-d(0)}^0 \eta_4^T(s)Q_1\eta_4(s)ds \\ &+ \int_{-h}^{-d(0)} \eta_5^T(s)Q_2\eta_5(s)ds + h\int_{-h}^0\int_\theta^0 \dot{x}^T(s)W\dot{x}(s)dsd\theta + \frac{h^2}{2}\int_{-h}^0\int_\sigma^0\int_\theta^0 \dot{x}^T(s)Z\dot{x}(s)dsd\theta d\sigma \\ &\leq [\lambda_{max}(\bar{P}_{11}) + \lambda_{max}(\bar{P}_{22}) + \lambda_{max}(\bar{P}_{33}) + h^2\lambda_{max}(\bar{P}_{44}) + h^2\lambda_{max}(\bar{P}_{55}) + 2\lambda_{max}(\bar{P}_{12}) \\ &+ 2\lambda_{max}(\bar{P}_{13}) + 2h\lambda_{max}(\bar{P}_{14}) + 2h\lambda_{max}(\bar{P}_{15}) + 2\lambda_{max}(\bar{P}_{23}) + 2h\lambda_{max}(\bar{P}_{24}) + 2h\lambda_{max}(\bar{P}_{25}) \end{aligned}$$

$$
\begin{aligned}
&+2h\lambda_{max}(\bar{P}_{34})+2h\lambda_{max}(\bar{P}_{35})+2h^2\lambda_{max}(\bar{P}_{45})+h\lambda_{max}(\bar{S}_{11})+h\lambda_{max}(\bar{S}_{22})+2h\lambda_{max}(\bar{S}_{12})\\
&+h\lambda_{max}(\bar{s}_{11})+h\lambda_{max}(\bar{s}_{22})+2h\lambda_{max}(\bar{s}_{12})+h\lambda_{max}(\bar{Q}_{11})+h\lambda_{max}(\bar{Q}_{22})+\frac{h^3}{3}\lambda_{max}(\bar{Q}_{33})\\
&+2h\lambda_{max}(\bar{Q}_{12})+h^2\lambda_{max}(\bar{Q}_{13})+h^2\lambda_{max}(\bar{Q}_{23})+h\lambda_{max}(\bar{q}_{11})+h\lambda_{max}(\bar{q}_{22})+\frac{h^3}{3}\lambda_{max}(\bar{q}_{33})\\
&+2h\lambda_{max}(\bar{q}_{12})+h^2\lambda_{max}(\bar{q}_{13})+h^2\lambda_{max}(\bar{q}_{23})+\frac{h^3}{2}\lambda_{max}(\bar{W})+\frac{h^5}{12}\lambda_{max}(\bar{Z})]\\
&\times sup_{-h\le\theta\le 0}\{x^T(\theta)Rx(\theta),\dot{x}^T(\theta)R\dot{x}(\theta)\}\\
\le&[\lambda_1+\lambda_2+\lambda_3+h^2(\lambda_4+\lambda_5+\lambda_{26}+\lambda_{27}+\lambda_{32}+\lambda_{33})+2(\lambda_6+\lambda_7+\lambda_{10})+2h(\lambda_8+\lambda_9\\
&+\lambda_{11}+\lambda_{12}+\lambda_{13}+\lambda_{14}+\lambda_{18}+\lambda_{21}+\lambda_{25}+\lambda_{31})+2h^2\lambda_{15}+h(\lambda_{16}+\lambda_{17}+\lambda_{19}+\lambda_{20}\\
&+\lambda_{22}+\lambda_{23}+\lambda_{28}+\lambda_{29})+\frac{h^3}{3}(\lambda_{24}+\lambda_{30})+\frac{h^3}{2}\lambda_{34}+\frac{h^5}{12}\lambda_{35}]c_1.
\end{aligned}
$$

We substitute the above two inequalities into (29) and assume the inequality (10) holds, we can easily derive that $x^T(t)Rx(t)\le c_2$ for all $t\in[0,T_f]$. Thus, the proof is completed. □

Remark 1. *The novel augmented Lyapunov–Krasovskii functional constructed in (11) takes advantage of information regarding the time-varying delay, which can make the obtained new finite-time boundedness condition less conservative. In addition, the Lyapunov–Krasovskii functional (11) is more general due to the introduction of several augmented vectors and two delay-product-type terms, such as* $\eta_1(t)$, $\eta_4(s)$, $\eta_5(s)$, $d(t)\eta_2^T(t)S_1\eta_2(t)$ *and* $(h-d(t))\eta_3^T(t)S_2\eta_3(t)$. *When several subblocks of the partitioned matrices* P,S_1,S_2,Q_1,Q_2 *are zero matrices with appropriate dimensions and* $W=0$, $Z=0$, *the augmented Lyapunov–Krasovskii functional* $V(x(t))$ *reduces to the simpler Lyapunov functions in some literature [24,26,34]. Additionally, to the best of our knowledge, the chosen Lyapunov–Krasovskii functional is a simple LKF instead of an augmented LKF in most existing studies regarding finite-time boundedness of dynamical systems, which is because the augmented LKF increases the difficulty of deriving finite-time boundedness criteria in terms of LMIs. However, this problem has been successfully solved in Theorem 1.*

Remark 2. *In Theorem 1, the improved reciprocally convex combination inequality and the auxiliary function-based integral inequalities are utilized to estimate the bound of the derivative of the constructed LKF. The auxiliary function-based integral inequalities are more general, as they can reduce to some other integral inequalities by appropriately choosing the auxiliary functions [23], such as the Jensen inequality, Bessel–Legendre inequality and Wirtinger-based integral inequality. In addition, the improved reciprocally convex combination inequality can provide a maximum lower bound with less slack matrix variables for several reciprocally convex combinations, which plays a critical role in reducing the conservativeness and the calculation complexity of the delay-dependent finite-time boundedness conditions for T-S fuzzy systems with time-varying delay and norm-bounded disturbance.*

3.2. Controller Design

Based on the delay-dependent finite-time boundedness criterion proposed in Theorem 1, we develop a memory state feedback controller to ensure the finite-time boundness of the resulting closed-loop T-S fuzzy time-delay system in the following theorem, which can be derived by solving a feasibility problem in terms of the linear matrix inequalities.

Theorem 2. *For the given scalars* $h>0$, μ_1, *and* μ_2, *the T-S fuzzy system (5) with a time-varying delay* $d(t)$ *satisfying (1) is finite-time bounded with respect to* (c_1,c_2,T_f,R,δ,h), *if there exist scalars* $\beta>0$, γ, *symmetric positive definite matrices* $\tilde{P}\in\boldsymbol{R}^{5n\times 5n}$, $\tilde{S}_1,\tilde{S}_2\in\boldsymbol{R}^{2n\times 2n}$, $\tilde{Q}_1,\tilde{Q}_2\in\boldsymbol{R}^{3n\times 3n}$, $\tilde{W},\tilde{Z},\tilde{U}\in\boldsymbol{R}^{n\times n}$, *any matrices* $\tilde{Y}_1,\tilde{Y}_2\in\boldsymbol{R}^{3n\times 3n}$, $X\in\boldsymbol{R}^{n\times n}$ *and* $L_{1j},L_{2j}\in\boldsymbol{R}^{p\times n}(j=1,2,\ldots,r)$, *such that the following conditions hold:*

$$\begin{pmatrix} \tilde{\Sigma}_{ii}(0,\mu_1) & \tilde{\Lambda}_1^T\tilde{Y}_1 \\ \tilde{Y}_1^T\tilde{\Lambda}_1 & -\tilde{W}_0 \end{pmatrix} < 0, \quad i = 1,2,\ldots,r \tag{30}$$

$$\begin{pmatrix} \tilde{\Sigma}_{ii}(0,\mu_2) & \tilde{\Lambda}_1^T\tilde{Y}_1 \\ \tilde{Y}_1^T\tilde{\Lambda}_1 & -\tilde{W}_0 \end{pmatrix} < 0, \quad i = 1,2,\ldots,r \tag{31}$$

$$\begin{pmatrix} \tilde{\Sigma}_{ii}(h,\mu_1) & \tilde{\Lambda}_2^T\tilde{Y}_2^T \\ \tilde{Y}_2\tilde{\Lambda}_2 & -\tilde{W}_0 \end{pmatrix} < 0, \quad i = 1,2,\ldots,r \tag{32}$$

$$\begin{pmatrix} \tilde{\Sigma}_{ii}(h,\mu_2) & \tilde{\Lambda}_2^T\tilde{Y}_2^T \\ \tilde{Y}_2\tilde{\Lambda}_2 & -\tilde{W}_0 \end{pmatrix} < 0, \quad i = 1,2,\ldots,r \tag{33}$$

$$\begin{pmatrix} \tilde{\Psi}_{ij}(0,\mu_1) & \tilde{\Lambda}_1^T\tilde{Y}_1 \\ \tilde{Y}_1^T\tilde{\Lambda}_1 & \frac{1-r}{r}\tilde{W}_0 \end{pmatrix} < 0, \quad i,j = 1,2,\ldots,r,\ i \neq j \tag{34}$$

$$\begin{pmatrix} \tilde{\Psi}_{ij}(0,\mu_2) & \tilde{\Lambda}_1^T\tilde{Y}_1 \\ \tilde{Y}_1^T\tilde{\Lambda}_1 & \frac{1-r}{r}\tilde{W}_0 \end{pmatrix} < 0, \quad i,j = 1,2,\ldots,r,\ i \neq j \tag{35}$$

$$\begin{pmatrix} \tilde{\Psi}_{ij}(h,\mu_1) & \tilde{\Lambda}_2^T\tilde{Y}_2^T \\ \tilde{Y}_2\tilde{\Lambda}_2 & \frac{1-r}{r}\tilde{W}_0 \end{pmatrix} < 0, \quad i,j = 1,2,\ldots,r,\ i \neq j \tag{36}$$

$$\begin{pmatrix} \tilde{\Psi}_{ij}(h,\mu_2) & \tilde{\Lambda}_2^T\tilde{Y}_2^T \\ \tilde{Y}_2\tilde{\Lambda}_2 & \frac{1-r}{r}\tilde{W}_0 \end{pmatrix} < 0, \quad i,j = 1,2,\ldots,r,\ i \neq j \tag{37}$$

$$c_1\tilde{\Pi} + \tilde{\lambda}_{37}\delta < \tilde{\lambda}_{36}c_2e^{-\beta T_f}, \tag{38}$$

where

$$\begin{aligned} \tilde{P} &= \begin{pmatrix} \tilde{P}_{11} & \tilde{P}_{12} & \tilde{P}_{13} & \tilde{P}_{14} & \tilde{P}_{15} \\ * & \tilde{P}_{22} & \tilde{P}_{23} & \tilde{P}_{24} & \tilde{P}_{25} \\ * & * & \tilde{P}_{33} & \tilde{P}_{34} & \tilde{P}_{35} \\ * & * & * & \tilde{P}_{44} & \tilde{P}_{45} \\ * & * & * & * & \tilde{P}_{55} \end{pmatrix}, \tilde{Q}_1 = \begin{pmatrix} \tilde{Q}_{11} & \tilde{Q}_{12} & \tilde{Q}_{13} \\ * & \tilde{Q}_{22} & \tilde{Q}_{23} \\ * & * & \tilde{Q}_{33} \end{pmatrix}, \tilde{Q}_2 = \begin{pmatrix} \tilde{q}_{11} & \tilde{q}_{12} & \tilde{q}_{13} \\ * & \tilde{q}_{22} & \tilde{q}_{23} \\ * & * & \tilde{q}_{33} \end{pmatrix}, \\ \tilde{S}_1 &= \begin{pmatrix} \tilde{S}_{11} & \tilde{S}_{12} \\ * & \tilde{S}_{22} \end{pmatrix}, \ \tilde{S}_2 = \begin{pmatrix} \tilde{s}_{11} & \tilde{s}_{12} \\ * & \tilde{s}_{22} \end{pmatrix}, \end{aligned}$$

$$\begin{aligned} \tilde{\Sigma}_{ij}(d(t),\dot{d}(t)) =& Sym\{\tilde{\Xi}_1^T\tilde{P}\tilde{\Xi}_2\} + \dot{d}(t)\tilde{\Xi}_3^T\tilde{S}_1\tilde{\Xi}_3 - \dot{d}(t)\tilde{\Xi}_4^T\tilde{S}_2\tilde{\Xi}_4 + Sym(\tilde{\Xi}_3^T\tilde{S}_1\tilde{\Xi}_5 + \tilde{\Xi}_4^T\tilde{S}_2\tilde{\Xi}_6) \\ &+ Sym(\tilde{\Xi}_7^T\tilde{Q}_1\tilde{\Xi}_8) + \tilde{\Xi}_9^T\tilde{Q}_1\tilde{\Xi}_9 - (1-\dot{d}(t))\tilde{\Xi}_{10}^T\tilde{Q}_1\tilde{\Xi}_{10} + Sym(\tilde{\Xi}_{11}^T\tilde{Q}_2\tilde{\Xi}_{12}) \end{aligned}$$

$$
\begin{aligned}
&+ (1-\dot{d}(t))\tilde{\Xi}_{13}^T\tilde{Q}_2\tilde{\Xi}_{13} - \tilde{\Xi}_{14}^T\tilde{Q}_2\tilde{\Xi}_{14} + h^2\tilde{e}_{13}^T\tilde{W}\tilde{e}_{13} + \frac{h^4}{4}\tilde{e}_{13}^T\tilde{Z}\tilde{e}_{13} - h^2\tilde{\Xi}_{15}^T\tilde{Z}\tilde{\Xi}_{15} \\
&- 2h^2\tilde{\Xi}_{16}^T\tilde{Z}\tilde{\Xi}_{16} - \tilde{e}_{12}^T\tilde{U}\tilde{e}_{12} + (\alpha-2)\tilde{\Lambda}_1^T\tilde{W}_0\tilde{\Lambda}_1 - (\alpha+1)\tilde{\Lambda}_2^T\tilde{W}_0\tilde{\Lambda}_2 \\
&- Sym\{\tilde{\Lambda}_1^T[\alpha\tilde{Y}_1 + (1-\alpha)\tilde{Y}_2]\tilde{\Lambda}_2\} + Sym\{(\tilde{e}_1^T + \gamma\tilde{e}_{13}^T)[A_iX\tilde{e}_1 + B_iL_{1j}\tilde{e}_1 \\
&+ A_{di}X\tilde{e}_2 + B_iL_{2j}\tilde{e}_2 + G_iX\tilde{e}_{12} - X\tilde{e}_{13}]\},
\end{aligned}
$$

$$
\tilde{\Psi}_{ij}(d(t),\dot{d}(t)) = \frac{1}{r-1}\tilde{\Sigma}_{ii}(d(t),\dot{d}(t)) + \frac{1}{2}\tilde{\Sigma}_{ij}(d(t),\dot{d}(t)) + \frac{1}{2}\tilde{\Sigma}_{ji}(d(t),\dot{d}(t)),
$$

$$
\begin{aligned}
\alpha &= \frac{d(t)}{h}, \quad \tilde{W}_0 = diag\{\tilde{W}, 3\tilde{W}, 5\tilde{W}\}, \quad \tilde{\Xi}_1 = \begin{bmatrix}\tilde{e}_1^T & \tilde{e}_2^T & \tilde{e}_3^T & d(t)\tilde{e}_6^T & (h-d(t))\tilde{e}_7^T\end{bmatrix}^T, \\
\tilde{\Xi}_2 &= [\tilde{e}_{13}^T \quad (1-\dot{d}(t))\tilde{e}_4^T \quad \tilde{e}_5^T \quad \tilde{e}_1^T - (1-\dot{d}(t))\tilde{e}_2^T \quad (1-\dot{d}(t))\tilde{e}_2^T - \tilde{e}_3^T]^T, \\
\tilde{\Xi}_3 &= \begin{bmatrix}\tilde{e}_1^T & \tilde{e}_6^T\end{bmatrix}^T, \quad \tilde{\Xi}_4 = \begin{bmatrix}\tilde{e}_1^T & \tilde{e}_7^T\end{bmatrix}^T, \quad \tilde{\Xi}_5 = \begin{bmatrix}d(t)\tilde{e}_{13}^T & -\dot{d}(t)\tilde{e}_6^T + \tilde{e}_1^T - (1-\dot{d}(t))\tilde{e}_2^T\end{bmatrix}^T, \\
\tilde{\Xi}_6 &= \begin{bmatrix}(h-d(t))\tilde{e}_{13}^T & \dot{d}(t)\tilde{e}_7^T + (1-\dot{d}(t))\tilde{e}_2^T - \tilde{e}_3^T\end{bmatrix}^T, \quad \tilde{\Xi}_7 = \begin{bmatrix}d(t)\tilde{e}_6^T & \tilde{e}_1^T - \tilde{e}_2^T & d(t)(\tilde{e}_1^T - \tilde{e}_6^T)\end{bmatrix}^T, \\
\tilde{\Xi}_8 &= \begin{bmatrix}0 & 0 & \tilde{e}_{13}^T\end{bmatrix}^T, \quad \tilde{\Xi}_9 = \begin{bmatrix}\tilde{e}_1^T & \tilde{e}_{13}^T & 0\end{bmatrix}^T, \quad \tilde{\Xi}_{10} = \begin{bmatrix}\tilde{e}_2^T & \tilde{e}_4^T & \tilde{e}_1^T - \tilde{e}_2^T\end{bmatrix}^T, \\
\tilde{\Xi}_{11} &= [(h-d(t))\tilde{e}_7^T \quad \tilde{e}_2^T - \tilde{e}_3^T \quad (h-d(t))(\tilde{e}_2^T - \tilde{e}_7^T)]^T, \quad \tilde{\Xi}_{12} = \begin{bmatrix}0 & 0 & (1-\dot{d}(t))\tilde{e}_4^T\end{bmatrix}^T, \\
\tilde{\Xi}_{13} &= \begin{bmatrix}\tilde{e}_2^T & \tilde{e}_4^T & 0\end{bmatrix}^T, \quad \tilde{\Xi}_{14} = \begin{bmatrix}\tilde{e}_3^T & \tilde{e}_5^T & \tilde{e}_2^T - \tilde{e}_3^T\end{bmatrix}^T, \quad \tilde{\Xi}_{15} = \tilde{e}_1 - \tilde{e}_{10}, \quad \tilde{\Xi}_{16} = \tilde{e}_1 + 2\tilde{e}_{10} - 6\tilde{e}_{11}, \\
\tilde{\Lambda}_1 &= [\tilde{e}_1^T - \tilde{e}_2^T \quad \tilde{e}_1^T + \tilde{e}_2^T - 2\tilde{e}_6^T \quad \tilde{e}_1^T - \tilde{e}_2^T + 6\tilde{e}_6^T - 12\tilde{e}_8^T]^T, \\
\tilde{\Lambda}_2 &= [\tilde{e}_2^T - \tilde{e}_3^T \quad \tilde{e}_2^T + \tilde{e}_3^T - 2\tilde{e}_7^T \quad \tilde{e}_2^T - \tilde{e}_3^T + 6\tilde{e}_7^T - 12\tilde{e}_9^T]^T, \\
\tilde{\Pi} &= \tilde{\lambda}_1 + \tilde{\lambda}_2 + \tilde{\lambda}_3 + h^2(\tilde{\lambda}_4 + \tilde{\lambda}_5 + \tilde{\lambda}_{26} + \tilde{\lambda}_{27} + \tilde{\lambda}_{32} + \tilde{\lambda}_{33}) + 2(\tilde{\lambda}_6 + \tilde{\lambda}_7 + \tilde{\lambda}_{10}) \\
&\quad + 2h(\tilde{\lambda}_8 + \tilde{\lambda}_9 + \tilde{\lambda}_{11} + \tilde{\lambda}_{12} + \tilde{\lambda}_{13} + \tilde{\lambda}_{14} + \tilde{\lambda}_{18} + \tilde{\lambda}_{21} + \tilde{\lambda}_{25} + \tilde{\lambda}_{31}) + 2h^2\tilde{\lambda}_{15} \\
&\quad + h(\tilde{\lambda}_{16} + \tilde{\lambda}_{17} + \tilde{\lambda}_{19} + \tilde{\lambda}_{20} + \tilde{\lambda}_{22} + \tilde{\lambda}_{23} + \tilde{\lambda}_{28} + \tilde{\lambda}_{29}) + \frac{h^3}{3}(\tilde{\lambda}_{24} + \tilde{\lambda}_{30}) \\
&\quad + \frac{h^3}{2}\tilde{\lambda}_{34} + \frac{h^5}{12}\tilde{\lambda}_{35}, \\
\tilde{\lambda}_1 &= \lambda_{max}(\hat{P}_{11}), \quad \tilde{\lambda}_2 = \lambda_{max}(\hat{P}_{22}), \quad \tilde{\lambda}_3 = \lambda_{max}(\hat{P}_{33}), \quad \tilde{\lambda}_4 = \lambda_{max}(\hat{P}_{44}), \quad \tilde{\lambda}_5 = \lambda_{max}(\hat{P}_{55}), \\
\tilde{\lambda}_6 &= \lambda_{max}(\hat{P}_{12}), \quad \tilde{\lambda}_7 = \lambda_{max}(\hat{P}_{13}), \quad \tilde{\lambda}_8 = \lambda_{max}(\hat{P}_{14}), \quad \tilde{\lambda}_9 = \lambda_{max}(\hat{P}_{15}), \quad \tilde{\lambda}_{10} = \lambda_{max}(\hat{P}_{23}), \\
\tilde{\lambda}_{11} &= \lambda_{max}(\hat{P}_{24}), \quad \tilde{\lambda}_{12} = \lambda_{max}(\hat{P}_{25}), \quad \tilde{\lambda}_{13} = \lambda_{max}(\hat{P}_{34}), \quad \tilde{\lambda}_{14} = \lambda_{max}(\hat{P}_{35}), \quad \tilde{\lambda}_{15} = \lambda_{max}(\hat{P}_{45}), \\
\tilde{\lambda}_{16} &= \lambda_{max}(\hat{S}_{11}), \quad \tilde{\lambda}_{17} = \lambda_{max}(\hat{S}_{22}), \quad \tilde{\lambda}_{18} = \lambda_{max}(\hat{S}_{12}), \quad \tilde{\lambda}_{19} = \lambda_{max}(\hat{s}_{11}), \quad \tilde{\lambda}_{20} = \lambda_{max}(\hat{s}_{22}), \\
\tilde{\lambda}_{21} &= \lambda_{max}(\hat{s}_{12}), \quad \tilde{\lambda}_{22} = \lambda_{max}(\hat{Q}_{11}), \quad \tilde{\lambda}_{23} = \lambda_{max}(\hat{Q}_{22}), \quad \tilde{\lambda}_{24} = \lambda_{max}(\hat{Q}_{33}), \quad \tilde{\lambda}_{25} = \lambda_{max}(\hat{Q}_{12}), \\
\tilde{\lambda}_{26} &= \lambda_{max}(\hat{Q}_{13}), \quad \tilde{\lambda}_{27} = \lambda_{max}(\hat{Q}_{23}), \quad \tilde{\lambda}_{28} = \lambda_{max}(\hat{q}_{11}), \quad \tilde{\lambda}_{29} = \lambda_{max}(\hat{q}_{22}), \quad \tilde{\lambda}_{30} = \lambda_{max}(\hat{q}_{33}), \\
\tilde{\lambda}_{31} &= \lambda_{max}(\hat{q}_{12}), \quad \tilde{\lambda}_{32} = \lambda_{max}(\hat{q}_{13}), \quad \tilde{\lambda}_{33} = \lambda_{max}(\hat{q}_{23}), \quad \tilde{\lambda}_{34} = \lambda_{max}(\hat{W}), \quad \tilde{\lambda}_{35} = \lambda_{max}(\hat{Z}), \\
\tilde{\lambda}_{36} &= \lambda_{min}(\hat{P}_{11}), \quad \tilde{\lambda}_{37} = \lambda_{max}(\hat{U}), \\
\hat{P}_{1j} &= R^{-\frac{1}{2}}X\tilde{P}_{1j}X^{-1}R^{-\frac{1}{2}}, \ j=1,2,3,4,5, \quad \hat{P}_{2j} = R^{-\frac{1}{2}}X\tilde{P}_{2j}X^{-1}R^{-\frac{1}{2}}, \ j=2,3,4,5, \\
\hat{P}_{3j} &= R^{-\frac{1}{2}}X\tilde{P}_{3j}X^{-1}R^{-\frac{1}{2}}, \ j=3,4,5, \quad \hat{P}_{4j} = R^{-\frac{1}{2}}X\tilde{P}_{4j}X^{-1}R^{-\frac{1}{2}}, \ j=4,5, \quad \hat{P}_{55} = R^{-\frac{1}{2}}X\tilde{P}_{55}X^{-1}R^{-\frac{1}{2}}, \\
\hat{S}_{1j} &= R^{-\frac{1}{2}}X\tilde{S}_{1j}X^{-1}R^{-\frac{1}{2}}, \ j=1,2, \quad \hat{S}_{22} = R^{-\frac{1}{2}}X\tilde{S}_{22}X^{-1}R^{-\frac{1}{2}}, \quad \hat{s}_{1j} = R^{-\frac{1}{2}}X\tilde{s}_{1j}X^{-1}R^{-\frac{1}{2}}, \ j=1,2, \\
\hat{s}_{22} &= R^{-\frac{1}{2}}X\tilde{s}_{22}X^{-1}R^{-\frac{1}{2}}, \quad \hat{Q}_{1j} = R^{-\frac{1}{2}}X\tilde{Q}_{1j}X^{-1}R^{-\frac{1}{2}}, j=1,2,3, \quad \hat{Q}_{2j} = R^{-\frac{1}{2}}X\tilde{Q}_{2j}X^{-1}R^{-\frac{1}{2}}, j=2,3, \\
\hat{Q}_{33} &= R^{-\frac{1}{2}}X\tilde{Q}_{33}X^{-1}R^{-\frac{1}{2}}, \quad \hat{q}_{1j} = R^{-\frac{1}{2}}X\tilde{q}_{1j}X^{-1}R^{-\frac{1}{2}}, \ j=1,2,3, \quad \hat{q}_{2j} = R^{-\frac{1}{2}}X\tilde{q}_{2j}X^{-1}R^{-\frac{1}{2}}, \ j=2,3, \\
\hat{q}_{33} &= R^{-\frac{1}{2}}X\tilde{q}_{33}X^{-1}R^{-\frac{1}{2}}, \quad \hat{W} = R^{-\frac{1}{2}}X\tilde{W}X^{-1}R^{-\frac{1}{2}}, \quad \hat{Z} = R^{-\frac{1}{2}}X\tilde{Z}X^{-1}R^{-\frac{1}{2}}, \quad \hat{U} = X\tilde{U}X^{-1}.
\end{aligned}
$$

In this case, the memory state feedback controller gains are given by $K_{1j} = L_{1j}X^{-1}$, $K_{2j} = L_{2j}X^{-1}$, $j = 1, 2, \ldots, r$.

Proof. Choose the Lyapunov–Krasovskii functional candidate (11) again for the resulting closed-loop T-S fuzzy time-delay system (5).

From the proof of Theorem 1, we obtain the inequality (22):

$$\dot{V}(x(t)) \leq \sum_{i=1}^{r} \rho_i(\xi(t))\varepsilon^T(t)\Sigma_i(d(t), \dot{d}(t))\varepsilon(t) + \omega^T(t)U\omega(t).$$

Furthermore, it can be easily obtained that

$$\dot{V}(x(t)) \leq \tilde{\varepsilon}^T(t)\hat{\Sigma}(d(t), \dot{d}(t))\tilde{\varepsilon}(t) + \omega^T(t)U\omega(t), \tag{39}$$

where $\hat{\Sigma}(d(t), \dot{d}(t)) = \hat{\Sigma}_1(d(t), \dot{d}(t)) + \hat{\Sigma}_2(d(t))$,

$$\begin{aligned}\hat{\Sigma}_1(d(t), \dot{d}(t)) =& Sym\{\tilde{\Xi}_1^T P\tilde{\Xi}_2\} + \dot{d}(t)\tilde{\Xi}_3^T S_1\tilde{\Xi}_3 - \dot{d}(t)\tilde{\Xi}_4^T S_2\tilde{\Xi}_4 + Sym(\tilde{\Xi}_3^T S_1\tilde{\Xi}_5 + \tilde{\Xi}_4^T S_2\tilde{\Xi}_6) \\ &+ Sym(\tilde{\Xi}_7^T Q_1\tilde{\Xi}_8) + \tilde{\Xi}_9^T Q_1\tilde{\Xi}_9 - (1 - \dot{d}(t))\tilde{\Xi}_{10}^T Q_1\tilde{\Xi}_{10} + Sym(\tilde{\Xi}_{11}^T Q_2\tilde{\Xi}_{12}) \\ &+ (1 - \dot{d}(t))\tilde{\Xi}_{13}^T Q_2\tilde{\Xi}_{13} - \tilde{\Xi}_{14}^T Q_2\tilde{\Xi}_{14} + h^2\tilde{e}_{13}^T W\tilde{e}_{13} + \frac{h^4}{4}\tilde{e}_{13}^T Z\tilde{e}_{13} - h^2\tilde{\Xi}_{15}^T Z\tilde{\Xi}_{15} \\ &- 2h^2\tilde{\Xi}_{16}^T Z\tilde{\Xi}_{16} - \tilde{e}_{12}^T U\tilde{e}_{12} + (\alpha - 2)\tilde{\Lambda}_1^T W_0\tilde{\Lambda}_1 - (\alpha + 1)\tilde{\Lambda}_2^T W_0\tilde{\Lambda}_2 \\ &- Sym\{\tilde{\Lambda}_1^T[\alpha Y_1 + (1 - \alpha)Y_2]\tilde{\Lambda}_2\},\end{aligned}$$

$$\hat{\Sigma}_2(d(t)) = (1 - \alpha)\tilde{\Lambda}_1^T Y_1 W_0^{-1} Y_1^T\tilde{\Lambda}_1 + \alpha\tilde{\Lambda}_2^T Y_2^T W_0^{-1} Y_2\tilde{\Lambda}_2.$$

According to the inequality (39), we have $\dot{V}(x(t)) < \beta V(x(t)) + \omega^T(t)U\omega(t)$, if the following inequality holds,

$$\tilde{\varepsilon}^T(t)\hat{\Sigma}(d(t), \dot{d}(t))\tilde{\varepsilon}(t) < 0. \tag{40}$$

Then, similar to Theorem 1, if the inequalities (10) and (40) hold, we can easily obtain that the closed-loop T-S fuzzy time-delay system (5) is finite-time bounded with respect to $(c_1, c_2, T_f, R, \delta, h)$.

Now, the closed-loop T-S fuzzy time-delay system (5) can be rewritten as:

$$\Theta(t)\tilde{\varepsilon}(t) = 0, \tag{41}$$

where $\Theta(t) = [A(t) + B(t)K_1(t) \quad A_d(t) + B(t)K_2(t) \quad 0 \quad 0 \quad 0 \quad 0 \quad 0 \quad 0 \quad 0 \quad 0 \quad 0 \quad G(t) \quad -I]$.

According to Finsler lemma, from (40) and (41), it can be obtained that the closed-loop T-S fuzzy time-delay system (5) is finite-time bounded with respect to $(c_1, c_2, T_f, R, \delta, h)$ if there exists a matrix $\Phi \in \mathbf{R}^{13n \times n}$, such that:

$$\hat{\Sigma}(d(t), \dot{d}(t)) + Sym\{\Phi\Theta(t)\} < 0. \tag{42}$$

Let $\Phi = [X^{-1} \quad 0 \quad 0 \quad 0 \quad 0 \quad 0 \quad 0 \quad 0 \quad 0 \quad 0 \quad 0 \quad 0 \quad \gamma X^{-1}]^T$, where γ is an arbitrary scalar. Then, we have the inequality (42) is equivalent to

$$\begin{aligned}&\hat{\Sigma}(d(t), \dot{d}(t)) + Sym\{(\tilde{e}_1^T X^{-T} + \gamma\tilde{e}_{13}^T X^{-T})[A(t)\tilde{e}_1 + B(t)K_1(t)\tilde{e}_1 \\ &\quad + A_d(t)\tilde{e}_2 + B(t)K_2(t)\tilde{e}_2 + G(t)\tilde{e}_{12} - \tilde{e}_{13}]\} < 0.\end{aligned} \tag{43}$$

Let $\Gamma_1 = diag(X, X, X, X, X, X, X, X, X, X, X, X, X)$. Multiplying (43) left by Γ_1^T and right by Γ_1, we can obtain the equivalent condition of (43) as follows:

$$\Gamma_1^T \hat{\Sigma}(d(t),\dot{d}(t))\Gamma_1 + Sym\{(\tilde{e}_1^T + \gamma\tilde{e}_{13}^T)[A(t)X\tilde{e}_1 + B(t)K_1(t)X\tilde{e}_1 + A_d(t)X\tilde{e}_2 + B(t)K_2(t)X\tilde{e}_2 + G(t)X\tilde{e}_{12} - X\tilde{e}_{13}]\} < 0.$$

Let $\Gamma_2 = diag(X,X,X,X,X)$, $\Gamma_3 = diag(X,X)$, $\Gamma_4 = diag(X,X,X)$, $\tilde{P} = \Gamma_2^T P\Gamma_2$, $\tilde{S}_1 = \Gamma_3^T S_1\Gamma_3$, $\tilde{S}_2 = \Gamma_3^T S_2\Gamma_3$, $\tilde{Q}_1 = \Gamma_4^T Q_1\Gamma_4$, $\tilde{Q}_2 = \Gamma_4^T Q_2\Gamma_4$, $\tilde{W} = X^TWX$, $\tilde{Z} = X^TZX$, $\tilde{U} = X^TUX$, $\tilde{W}_0 = \Gamma_4^T W_0\Gamma_4$, $\tilde{Y}_1 = \Gamma_4^T Y_1\Gamma_4$, $\tilde{Y}_2 = \Gamma_4^T Y_2\Gamma_4$. It can be easily derived that $\Gamma_1^T\hat{\Sigma}(d(t),\dot{d}(t))\Gamma_1 = \tilde{\Sigma}_1(d(t),\dot{d}(t)) + \tilde{\Sigma}_2(d(t))$, where

$$\begin{aligned}\tilde{\Sigma}_1(d(t),\dot{d}(t)) =& Sym\{\tilde{\Xi}_1^T\tilde{P}\tilde{\Xi}_2\} + \dot{d}(t)\tilde{\Xi}_3^T\tilde{S}_1\tilde{\Xi}_3 - \dot{d}(t)\tilde{\Xi}_4^T\tilde{S}_2\tilde{\Xi}_4 + Sym(\tilde{\Xi}_3^T\tilde{S}_1\tilde{\Xi}_5 + \tilde{\Xi}_4^T\tilde{S}_2\tilde{\Xi}_6)\\ &+ Sym(\tilde{\Xi}_7^T\tilde{Q}_1\tilde{\Xi}_8) + \tilde{\Xi}_9^T\tilde{Q}_1\tilde{\Xi}_9 - (1-\dot{d}(t))\tilde{\Xi}_{10}^T\tilde{Q}_1\tilde{\Xi}_{10} + Sym(\tilde{\Xi}_{11}^T\tilde{Q}_2\tilde{\Xi}_{12})\\ &+ (1-\dot{d}(t))\tilde{\Xi}_{13}^T\tilde{Q}_2\tilde{\Xi}_{13} - \tilde{\Xi}_{14}^T\tilde{Q}_2\tilde{\Xi}_{14} + h^2\tilde{e}_{13}^T\tilde{W}\tilde{e}_{13} + \frac{h^4}{4}\tilde{e}_{13}^T\tilde{Z}\tilde{e}_{13} - h^2\tilde{\Xi}_{15}^T\tilde{Z}\tilde{\Xi}_{15}\\ &- 2h^2\tilde{\Xi}_{16}^T\tilde{Z}\tilde{\Xi}_{16} - \tilde{e}_{12}^T\tilde{U}\tilde{e}_{12} + (\alpha-2)\tilde{\Lambda}_1^T\tilde{W}_0\tilde{\Lambda}_1 - (\alpha+1)\tilde{\Lambda}_2^T\tilde{W}_0\tilde{\Lambda}_2\\ &- Sym\{\tilde{\Lambda}_1^T[\alpha\tilde{Y}_1 + (1-\alpha)\tilde{Y}_2]\tilde{\Lambda}_2\},\end{aligned}$$

$$\tilde{\Sigma}_2(d(t)) = (1-\alpha)\tilde{\Lambda}_1^T\tilde{Y}_1\tilde{W}_0^{-1}\tilde{Y}_1^T\tilde{\Lambda}_1 + \alpha\tilde{\Lambda}_2^T\tilde{Y}_2^T\tilde{W}_0^{-1}\tilde{Y}_2\tilde{\Lambda}_2.$$

Now, let $L_{1j} = K_{1j}X, L_{2j} = K_{2j}X, j = 1,2,\ldots,r$, then we can easily derive that (42) is equivalent to $\tilde{\Sigma}_{ij}(d(t),\dot{d}(t)) + \tilde{\Sigma}_2(d(t)) < 0$, where $\tilde{\Sigma}_{ij}(d(t),\dot{d}(t)) = \tilde{\Sigma}_1(d(t),\dot{d}(t)) + Sym\{(\tilde{e}_1^T + \gamma\tilde{e}_{13}^T)[A_iX\tilde{e}_1 + B_iL_{1j}\tilde{e}_1 + A_{di}X\tilde{e}_2 + B_iL_{2j}\tilde{e}_2 + G_iX\tilde{e}_{12} - X\tilde{e}_{13}]\}$, $i,j = 1,2,\ldots,r$. Additionally, it is clear that the condition (38) is the equivalent condition of (10).

According to the Schur complement lemma and Lemma 2 in [35], similar to the proof of Theorem 1, the conditions (30)–(38) can ensure the closed-loop T-S fuzzy time-delay system (5) finite-time bounded with respect to $(c_1, c_2, T_f, R, \delta, h)$, and we can obtain the memory state feedback controller gains $K_{1j} = L_{1j}X^{-1}, K_{2j} = L_{2j}X^{-1}, j = 1,2,\ldots,r$. Thus, this completes the proof of the theorem. □

Remark 3. *It is well known that the concept of finite-time boundedness reduces to the concept of finite-time stability when $\omega(t) = 0$. Thus, finite-time stability is a special case of finite-time boundedness. The authors in [28] discuss the problem of finite-time stability and stabilization for a class of T-S fuzzy systems with time-varying delay. However, this paper is concerned with finite-time boundness analysis and the finite-time stabilization problem for T-S fuzzy systems with a time-varying delay and norm-bounded disturbance. Therefore, the developed results in this paper are more general.*

Remark 4. *The Finsler lemma is employed to design the memory state feedback controller for a T-S fuzzy time-delay system in the proof of Theorem 2. In order to derive a finite-time stabilization condition in the form of LMIs, the matrix $\Phi = [X^{-1}\ \ 0\ \ 0\ \ 0\ \ 0\ \ 0\ \ 0\ \ 0\ \ 0\ \ 0\ \ 0\ \ 0\ \ \gamma X^{-1}]^T$ is defined, which may introduce some conservativeness. However, it should be mentioned that this is indeed an effective approach to obtain a finite-time stabilization criterion. To reduce the aforementioned conservativeness, an appropriate parameter γ can be obtained by applying several powerful optimization algorithms.*

Remark 5. *Based on the parallel distributed compensation scheme, the memory state feedback controller is designed to ensure finite-time boundness of the corresponding closed-loop T-S fuzzy time-delay system in Theorem 2. For fixed $(c_1, c_2, T_f, R, \delta, h)$, the optimal minimum values of c_2 for guaranteeing the closed-loop T-S fuzzy system finite-time bounded can be obtained by solving a series of LMIs, namely,*

$$\min_{(30)-(38)} c_2.$$

The memory state feedback controller gains are given by $K_{1j} = L_{1j}X^{-1}, K_{2j} = L_{2j}X^{-1}, j = 1,2,\ldots,r$.

4. Numerical Example

In this section, a numerical example is given to illustrate the effectiveness of the proposed results.

This example deals with a truck-trailer system with time-varying delay. The dynamic model is described as follows:

$$\begin{cases} \dot{x}_1(t) = -a\frac{v\bar{t}}{Lt_0}x_1(t) - (1-a)\frac{v\bar{t}}{Lt_0}x_1(t-d(t)) + \frac{v\bar{t}}{lt_0}u(t) + \omega_1(t) \\ \dot{x}_2(t) = a\frac{v\bar{t}}{Lt_0}x_1(t) + (1-a)\frac{v\bar{t}}{Lt_0}x_1(t-d(t)) \\ \dot{x}_3(t) = \frac{v\bar{t}}{t_0}\sin[x_2(t) + a\frac{v\bar{t}}{L}x_1(t) + (1-a)\frac{v\bar{t}}{2L}x_1(t-d(t))] \end{cases}$$

where $x_1(t)$ is the angle difference between the truck and the trailer, $x_2(t)$ is the angle of the trailer, $x_3(t)$ represents the vertical position of the rear end of the trailer, $u(t)$ denotes the steering angle, $\omega(t) = \left(\omega_1^T(t)\ \omega_2^T(t)\ \omega_3^T(t)\right)^T$ is the exogenous disturbance.

Let $\sigma(t) = x_2(t) + a\frac{v\bar{t}}{L}x_1(t) + (1-a)\frac{v\bar{t}}{2L}x_1(t-d(t))$, the T-S fuzzy time-delay system that represents the above truck-tailer model is as follows:

Plant Rule 1: If $\sigma(t)$ is about 0, then

$$\dot{x}(t) = A_1x(t) + A_{d1}x(t-d(t)) + B_1u(t) + B_{\omega 1}\omega(t);$$

Plant Rule 2: If $\sigma(t)$ is about $\pm\pi$, then

$$\dot{x}(t) = A_2x(t) + A_{d2}x(t-d(t)) + B_2u(t) + B_{\omega 2}\omega(t),$$

where

$$A_1 = \begin{pmatrix} -a\frac{v\bar{t}}{Lt_0} & 0 & 0 \\ a\frac{v\bar{t}}{Lt_0} & 0 & 0 \\ a\frac{v^2\bar{t}^2}{2Lt_0} & \frac{v\bar{t}}{t_0} & 0 \end{pmatrix},\ A_{d1} = \begin{pmatrix} -b\frac{v\bar{t}}{Lt_0} & 0 & 0 \\ b\frac{v\bar{t}}{Lt_0} & 0 & 0 \\ b\frac{v^2\bar{t}^2}{2Lt_0} & 0 & 0 \end{pmatrix},$$

$$A_2 = \begin{pmatrix} -a\frac{v\bar{t}}{Lt_0} & 0 & 0 \\ a\frac{v\bar{t}}{Lt_0} & 0 & 0 \\ a\frac{dv^2\bar{t}^2}{2Lt_0} & \frac{dv\bar{t}}{t_0} & 0 \end{pmatrix},\ A_{d2} = \begin{pmatrix} -b\frac{v\bar{t}}{Lt_0} & 0 & 0 \\ b\frac{v\bar{t}}{Lt_0} & 0 & 0 \\ b\frac{dv^2\bar{t}^2}{2Lt_0} & 0 & 0 \end{pmatrix},$$

$$B_1 = B_2 = \begin{pmatrix} \frac{v\bar{t}}{lt_0} \\ 0 \\ 0 \end{pmatrix},\quad B_{\omega 1} = B_{\omega 2} = \begin{pmatrix} 1 & 0 & 0 \\ 0 & 0 & 0 \\ 0 & 0 & 0 \end{pmatrix},$$

with $a + b = 1$.

In order to illustrate the developed results, we borrow the model parameters from [36], such as $a = 0.7$, $v = -1.0$, $L = 5.5$, $l = 2.8$, $\bar{t} = 2.0$, $t_0 = 0.5$, and $d = 10t_0/\pi$. The membership functions are defined as $\rho_1(x(t)) = 1/\left(1 + exp(x_1(t) + 0.5)\right)$, $\rho_2(x(t)) = 1 - \rho_1(x(t))$. Additionally, the other parameters involved in the simulation are chosen as $c_1 = 1$, $\delta = 0.3$, $\beta = 0.01$, $\gamma = 0.8$, $T_f = 10$, $R = I$, $\mu_1 = -0.1$, $\mu_2 = 0.1$, and $h = 0.6$. We aim to design a memory state feedback controller such that the resulting closed-loop T-S fuzzy time-delay system is finite-time bounded. By solving the LMI-based finite-time stabilization criterion proposed in Theorem 2 using the Matlab LMI toolbox, we can derive the feasible solutions for the optimal minimum value of $c_2 = 2.8830$. Furthermore, all the control gain matrices are obtained as follows:

$$K_{11} = L_{11}X^{-1} = \begin{pmatrix} 6.9875 & -13.5452 & 1.4041 \end{pmatrix},\ K_{12} = L_{12}X^{-1} = \begin{pmatrix} 6.9871 & -13.7540 & 1.3966 \end{pmatrix},$$

$$K_{21} = L_{21}X^{-1} = \begin{pmatrix} 0.3679 & 0.0085 & -0.0010 \end{pmatrix}, \quad K_{22} = L_{22}X^{-1} = \begin{pmatrix} 0.3861 & -0.0013 & 0.0001 \end{pmatrix}.$$

For the simulation framework, the exogenous disturbance is selected as $\omega(t) = (0.06 \sin t \;\; 0 \;\; 0)^T$, and the time-varying delay is assumed to be $d(t) = 0.25 + 0.25 \sin(0.3t)$. For the initial condition $x(0) = (0.8 \;\; -0.5 \;\; 0.2)^T$, the state response of the corresponding closed-loop T-S fuzzy time-delay system is depicted in Figure 1, and the evolution of $x^T(t)Rx(t)$ is shown in Figure 2. From the simulation results, it is obvious that the closed-loop T-S fuzzy time-delay system is finite-time bounded with respect to $(1, 2.8830, 10, I, 0.3, 0.6)$ via the above memory state feedback controller. In addition, for different h, the optimal minimum values of c_2 for ensuring the closed-loop T-S fuzzy system finite-time bounded are summarized in Table 1. This proves the effectiveness of our developed results in Theorem 2.

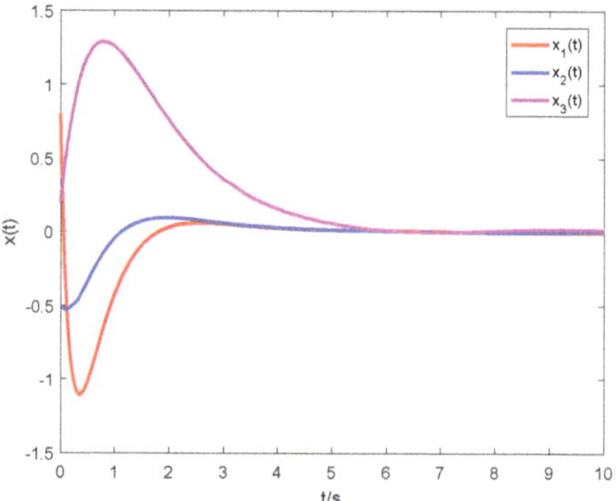

Figure 1. The state response of the closed-loop Takagi–Sugeno fuzzy system.

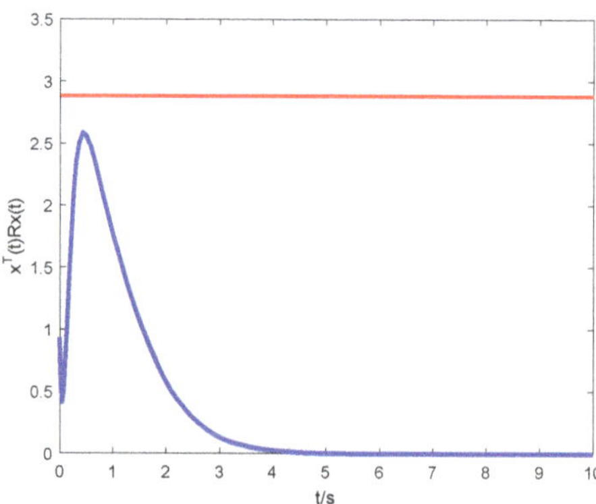

Figure 2. The time history of $x^T(t)Rx(t)$.

Table 1. The optimum bound values of c_2 for different h.

h	0.6	0.8	1.0	1.2	1.4
c_2	2.8830	3.6111	4.5002	5.7103	6.7601

5. Conclusions

In this paper, the problem of finite-time boundedness and finite-time stabilization for a class of T-S fuzzy time-delay systems was discussed. First, based on a new augmented LKF and by applying an improved reciprocally convex combination technique, a novel delay-dependent finite-time boundedness sufficient condition has been derived for an open-loop T-S fuzzy time-delay system. Secondly, a memory state feedback controller has been developed to ensure the finite-time boundedness of the corresponding closed-loop T-S fuzzy time-delay system. Finally, the effectiveness and advantages of the presented methods were demonstrated by a numerical example. Our future research work will focus on the problem of robust finite-time control for uncertain T-S fuzzy systems with time-varying delay and exogenous disturbance.

Author Contributions: Conceptualization, Y.R. and T.H.; methodology, Y.R.; software, Y.R.; validation, Y.R. and T.H.; formal analysis, Y.R.; investigation, Y.R. and T.H.; resources, T.H.; writing—original draft preparation, Y.R.; writing—review and editing, T.H.; visualization, Y.R. and T.H.; supervision, T.H. All authors have read and agreed to the published version of the manuscript.

Funding: This research was supported by the National Natural Science Foundation of China under Grant 61372187, Grant 61473239, and Grant 61702317.

Conflicts of Interest: The authors declare no conflict of interest.

References

1. Wang, H.Q.; Liu, P.X.; Li, S.; Wang, D. Adaptive neural output-feedback control for a class of nonlower triangular nonlinear systems with unmodeled dynamics. *IEEE Trans. Neural Netw. Learn. Syst.* **2018**, *29*, 3658–3668.
2. Zhao, X.D.; Wang, X.Y.; Zhang, S.; Zong, G.D. Adaptive neural backstepping control design for a class of nonsmooth nonlinear systems. *IEEE Trans. Syst. Man Cybern. Syst.* **2019**, *49*, 1820–1831. [CrossRef]
3. Roy, S.; Kar, I.N. Adaptive sliding mode control of a class of nonlinear systems with artificial delay. *J. Frankl. Inst.* **2017**, *354*, 8156–8179. [CrossRef]
4. Qiu, J.B.; Sun, K.K.; Wang, T.; Gao, H.J. Observer-based fuzzy adaptive event-triggered control for pure-feedback nonlinear systems with prescribed performance. *IEEE Trans. Fuzzy Syst.* **2019**, *27*, 2152–2162. [CrossRef]
5. Vrabel, R. Eigenvalue based approach for assessment of global robustness of nonlinear dynamical systems. *Symmetry* **2019**, *11*, 569. [CrossRef]
6. Li, X.; Zhu, Z.C.; Rui, G.C.; Cheng, D.; Shen, G.; Tang, Y. Force loading tracking control of an electro-hydraulic actuator based on a nonlinear adaptive fuzzy backstepping control scheme. *Symmetry* **2018**, *10*, 155. [CrossRef]
7. Takagi, T.; Sugeno, M. Fuzzy identification of systems and its applications to modeling and control. *IEEE Trans. Syst. Man Cybern.* **1985**, *15*, 116–132. [CrossRef]
8. Zheng, W.; Wang, H.B.; Wang, H.R.; Wen, S.H. Stability analysis and dynamic output feedback controller design of T-S fuzzy systems with time-varying delays and external disturbances. *J. Comput. Appl. Math.* **2019**, *358*, 111–135. [CrossRef]
9. Tan, J.Y.; Dian, S.Y.; Zhao, T.; Chen, L. Stability and stabilization of T-S fuzzy systems with time delay via Wirtinger-based double integral inequality. *Neurocomputing* **2018**, *275*, 1063–1071. [CrossRef]
10. Seuret, A.; Gouaisbaut, F. Wirtinger-based integral inequality: Application to time-delay systems. *Automatica* **2013**, *49*, 2860–2866. [CrossRef]
11. Zhao, T.; Huang, M.B.; Dian, S.Y. Stability and stabilization of T-S fuzzy systems with two additive time-varying delays. *Inform. Sci.* **2019**, *494*, 174–192. [CrossRef]
12. Benzaouia, A.; Hajjaji, A.E. Conditions of stabilization of positive continuous Takagi-Sugeno fuzzy systems with delay. *Int. J. Fuzzy Syst.* **2018**, *20*, 750–758. [CrossRef]
13. Lian, Z.; He, Y.; Zhang, C.K.; Wu, M. Further robust stability analysis for uncertain Takagi-Sugeno fuzzy systems with time-varying delay via relaxed integral inequality. *Inform. Sci.* **2017**, *409–410*, 139–150. [CrossRef]
14. Zhao, T.; Dian, S.Y. State feedback control for interval type-2 fuzzy systems with time-varying delay and unreliable communication links. *IEEE Trans. Fuzzy Syst.* **2018**, *26*, 951–966. [CrossRef]
15. Liu, C.; Mao, X.; Xu, X.Z.; Zhang, H.B. Stability analysis of discrete-time switched T-S fuzzy systems with all subsystems unstable. *IEEE Access* **2019**, *7*, 50412–50418. [CrossRef]

16. An, J.Y.; Lin, W.G. Improved stability criteria for time-varying delayed T-S fuzzy systems via delay partitioning approach. *Fuzzy Sets Syst.* **2011**, *185*, 83–94. [CrossRef]
17. Yang, J.; Luo, W.P.; Wang, Y.H.; Duan, C.S. Improved stability criteria for T-S fuzzy systems with time-varying delay by delay-partitioning approach. *Int. J. Control Autom. Syst.* **2015**, *13*, 1521–1529. [CrossRef]
18. Zeng, H.B.; Park, J.H.; Xia, J.W.; Xiao, S.P. Improved delay-dependent stability criteria for T-S fuzzy systems with time-varying delay. *Appl. Math. Comput.* **2014**, *235*, 492–501. [CrossRef]
19. Kwon, O.M.; Park, M.J.; Lee, S.M.; Park, J.H. Augmented Lyapunov–Krasovskii functional approaches to robust stability criteria for uncertain Takagi-Sugeno fuzzy systems with time-varying delay. *Fuzzy Sets Syst.* **2012**, *201*, 1–19. [CrossRef]
20. Wu, M.; He, Y.; She, J.H.; Liu, G.P. Delay-dependent criteria for robust stability of time-varying delay systems. *Automatica* **2004**, *40*, 1435–1439. [CrossRef]
21. Gu, K.; Kharitonov, V.L.; Chen, J. *Stability of Time-Delay Systems*; Birkhauser: Basel, Switzerland, 2003.
22. Park, P.G.; Ko, J.W.; Jeong, C. Reciprocally convex approach to stability of systems with time-varying delays. *Automatica* **2011**, *47*, 235–238. [CrossRef]
23. Park, P.; Lee, W.; Lee, S.Y. Auxiliary function-based integral inequalities for quadratic functions and their applications to time-delay systems. *J. Frankl. Inst.* **2015**, *352*, 1378–1396. [CrossRef]
24. Zeng, H.B.; He, Y.; Wu, M.; She, J.H. Free-matrix-based integral inequality for stability analysis of systems with time-varying delay. *IEEE Trans. Autom. Control* **2015**, *60*, 2768–2772. [CrossRef]
25. Liu, F.; Wu, M.; He, Y.; Yokoyamab, R. New delay-dependent stability criteria for T-S fuzzy systems with time-varying delay. *Fuzzy Sets Syst.* **2010**, *161*, 2033–2042. [CrossRef]
26. Lian, Z.; He, Y.; Zhang, C.K.; Wu, M. Stability analysis for T-S fuzzy systems with time-varying delay via free-matrix-based integral inequality. *Int. J. Control Autom. Syst.* **2016**, *14*, 21–28. [CrossRef]
27. Amato, F.; Ariola, M.; Dorato, P. Finite-time control of linear systems subject to parametric uncertainties and disturbances. *Automatica* **2001**, *37*, 1459–1463. [CrossRef]
28. Liu, H.; Shi, P.; Karimi, H.R.; Chadli, M. Finite-time stability and stabilisation for a class of nonlinear systems with time-varying delay. *Int. J. Syst. Sci.* **2016**, *47*, 1433–1444. [CrossRef]
29. Huang, X.P.; Wu, C.Y.; Liu, Y.P. Finite-time H_∞ model reference control of SLPV systems and its application to aero-engines. *IEEE Access* **2019**, *7*, 43525–43533. [CrossRef]
30. Ma, R.C.; Jiang, B.; Liu, Y. Finite-time stabilization with output-constraints of a class of high-order nonlinear systems. *Int. J. Control Autom. Syst.* **2018**, *16*, 945–952. [CrossRef]
31. Sakthivel, R.; Saravanakumar, T.; Kaviarasan, B.; Lim, Y.D. Finite-time dissipative based fault-tolerant control of Takagi-Sugeno fuzzy systems in a network environment. *J. Frankl. Inst.* **2017**, *354*, 3430–3455. [CrossRef]
32. Ren, C.C.; Ai, Q.L.; He, S.P. Finite-time non-fragile control of a class of uncertain linear positive systems. *IEEE Access* **2019**, *7*, 6319–6326. [CrossRef]
33. Zhang, X.M.; Han, Q.L.; Seuret, A.; Gouaisbaut, F. An improved reciprocally convex inequality and an augmented Lyapunov–Krasovskii functional for stability of linear systems with time-varying delay. *Automatica* **2017**, *84*, 221–226. [CrossRef]
34. Cao, Y.Y.; Frank, P.M. Stability analysis and synthesis of nonlinear time-delay systems via linear Takagi-Sugeno fuzzy models. *Fuzzy Sets Syst.* **2001**, *124*, 213–229. [CrossRef]
35. Han, L.; Qiu, C.Y.; Xiao, J. Finite-time H_∞ control synthesis for nonlinear switched systems using T-S fuzzy model. *Neurocomputing* **2016**, *171*, 156–170. [CrossRef]
36. Yan, H.C.; Wang, T.T.; Zhang, H.; Shi, H.B. Event-triggered H_∞ control for uncertain networked T-S fuzzy systems with time delay. *Neurocomputing* **2015**, *157*, 273–279. [CrossRef]

Article

Analysis Exploring the Uniformity of Flow Distribution in Multi-Channels for the Application of Printed Circuit Heat Exchangers

Hanbing Ke [1], Yuansheng Lin [1], Zhiwu Ke [1], Qi Xiao [1], Zhiguo Wei [1], Kai Chen [1] and Huijin Xu [2,*]

1 Science and Technology on Thermal Energy and Power Laboratory, Wuhan Second Ship Design and Resource Institute, Wuhan 430205, China; kehanbingwork@163.com (H.K.); keylab_rndl@163.com (Y.L.); jchdlzhjs@126.com (Z.K.); zhiyan7@sina.com (Q.X.); za2002@163.com (Z.W.); xjtuchen@foxmail.com (K.C.)
2 China-UK Low Carbon College, Shanghai Jiao Tong University, Shanghai 200240, China
* Correspondence: xuhuijin@sjtu.edu.cn or hjxu1015@gmail.com; Tel.: +86-18561575028

Received: 22 December 2019; Accepted: 6 February 2020; Published: 22 February 2020

Abstract: The maldistribution of fluid flow through multi-channels is a critical issue encountered in many areas, such as multi-channel heat exchangers, electronic device cooling, refrigeration and cryogenic devices, air separation and the petrochemical industry. In this paper, the uniformity of flow distribution in a printed circuit heat exchanger (PCHE) is investigated. The flow distribution and resistance characteristics of a PCHE plate are studied with numerical models under different flow distribution cases. The results show that the sudden change in the angle of the fluid at the inlet of the channel can be greatly reduced by using a spreader plate with an equal inner and outer radius. The flow separation of the fluid at the inlet of the channel can also be weakened and the imbalance of flow distribution in the channel can be reduced. Therefore, the flow uniformity can be improved and the pressure loss between the inlet and outlet of PCHEs can be reduced. The flow maldistribution in each PCHE channel can be reduced to ± 0.2%, and the average flow maldistribution in all PCHE channels can be reduced to less than 5% when the number of manifolds reaches nine. The numerical simulation of fluid flow distribution can provide guidance for the subsequent research and the design and development of multi-channel heat exchangers. In summary, the symmetry of the fluid flow in multi-channels for PCHE was analyzed in this work. This work presents the frequently encountered problem of maldistribution of fluid flow in engineering, and the performance promotion leads to symmetrical aspects in both the structure and the physical process.

Keywords: flow distribution; maldistribution; numerical modeling; symmetry; printed circuit heat exchanger

---

1. Introduction

A heat exchanger is a device used to transfer heat from a hot fluid to a cold fluid to meet specified process requirements, which is an industrial application of convection heat transfer and heat conduction [1]. The heat exchanger is an important piece of ship power system equipment, because the efficiency and cost of the ship power system are both significantly affected by the thermal-hydraulic performance of the intermediate heat exchanger [2]. The ship power system has a large amount of heat generation, in which its internal high heat flux electronic equipment has a large demand for heat dissipation, so it needs to adopt efficient cooling and heat dissipation technology in a limited space. It is necessary to develop a highly efficient heat exchanger with a high temperature and high pressure for the ship power system.

Double-pipe heat exchangers are the simplest exchangers used in industry. On one hand, these heat exchangers are cheap for both design and maintenance, making them a good choice for small industries.

On the other hand, their low efficiency, coupled with the large amount of space they occupy in large scales, has led modern industries to use more efficient heat exchangers, such as shell-and-tube or plate [3]. Considering the special nature of the marine environment, the traditional shell-and-tube heat exchangers are widely used as ship heat exchangers, which highly satisfy the requirements of the power system [4]. However, with the development of ship power systems in the direction of integration, miniaturization and high reliability, the traditional shell-and-tube heat exchanger has gradually exposed the problems of its large volume, heavy weight and high safety risk in a long-term, high-pressure environment. So, the investigation of a new compact structure and technology that possesses a high reliability of heat transfer is an inevitable developmental trend in the ship power system [5].

As a new type of highly efficient compact heat exchanger, the printed circuit heat exchanger (PCHE) is composed of many micro-channel plates, which can be chemically etched to generate a variety of channels [6]. Following this, the hot and cold plates are alternately superimposed with diffusion welding to create a compact heat transfer unit, which has the advantages of high-pressure resistance, heat exchange efficiency and compactness [7].

Generally, the channel diameter of PCHE is about 0.1 to 2.0 mm and the density of the heat transfer area can be as high as 2500 m^2/m^3 [8]. The working conditions can be different depending on the range of applications. The maximum temperature can reach 900 °C, the working pressure can reach 60 MPa and the design life can even reach 30–60 years. Under the same heat transfer power, the volume of PCHE is only about 1/5 of the shell-and-tube heat exchangers [9]. Therefore, PCHE provides a new heat transfer method to achieve the future of the ship power system with characteristics of integration, miniaturization and high reliability.

However, due to geometry design features or operating conditions, flow maldistribution is a common problem that can significantly reduce the desired heat exchanger performance. Therefore, finding a way to improve flow uniformity in PCHEs, so as to reduce the heat exchanger size, design margins or achieve the desired production rate, is of vital importance [10].

The effects of non-uniform flow on the heat exchanger performance have been well investigated over the decades. Jiao et al. [11] proved that the performance of the flow distribution in the heat exchanger is effectively improved by the optimum design of the second header installation by experimental studies. The results indicate that the flow distribution becomes more uniform when the ratio of outlet pipe diameter to inlet pipe diameter of the two headers of the heat exchanger are equal. Bobbili et al. [12] carried out experimental investigations to find the flow and pressure difference across the port to the channel in plate heat exchangers for a wide range of Reynolds numbers. The results indicated that the flow maldistribution increases with the increasing overall pressure drop in the plate heat exchangers. Zhang et al. [13] used various distributor configurations with a plate-fin heat exchanger under different operating conditions to assess the resultant change in its flow distribution and thermal performance. The results showed that the effect of the inlet angle of distributor on the flow distribution is significant and flow maldistribution in the lateral and gross flow directions is different.

However, the experimental research has some shortcomings, such as the large number data errors and the failure to cover all heat exchanger conditions. Therefore, the numerical simulation of non-uniform flow is particularly important. Zhang et al. [14] used the CFD method to predict the fluid flow distribution in plate-fin heat exchangers, which was simulated according to the configuration of the plate-fin heat exchanger currently used in industry. Wen et al. [15] characterized the turbulent flow structure inside the entrance of the plate-fin heat exchanger by CFD simulation under similar conditions. The numerical results indicate that the performance of fluid maldistribution in conventional entrance deteriorated, while the improved configuration, with a punched baffle, can effectively improve the performance in both radial and axial direction. Sheik et al. [16] performed a numerical study of the flow patterns of compact plate-fin heat exchangers because air (gas) flow maldistribution in the headers affects the exchanger performance. Three typical compact plate-fin heat exchangers were analyzed using Fluent software for the quantification of the flow maldistribution effects with ideal and real cases in their study.

When studying the pressure and flow characteristics or the heat transfer in arbitrary fluid conditions, symmetry is a necessary consideration. Wei et al. [17] studied the pressure fluctuation and flow characteristics in a two-stage, double-suction centrifugal pump based on CFD analysis. Monitor points were arranged in the full flow channel from the inlet to the outlet of the pump in order to show all the pressure fluctuation characteristics of the pump. Due to some parts of the pump rotating, as well as the whole shape being axially symmetric, the monitor points were set to rotate with the impeller in the rotation parts. Afridi et al. [18] carried out an irreversibility analysis of hybrid nanofluid flow over a thin needle with the effects of energy dissipation. When conducting the theoretical study of the heat transfer and entropy generation in the flow of dissipative hybrid nanofluid, the coordinate system and the geometry of the physical flow model are only shown for half of the thin needle, because of its symmetry, which simplified the calculation process of this study. Therefore, it is necessary to analyze the symmetry of the heat exchanger, including its structure central symmetry, channel flow field symmetry and adjacent channel structure symmetry.

However, the multi-channel PCHE plate shows an obvious maldistribution effect in practical applications, which produces a great limitation in the performance promotion of the PCHE plate. This aspect was rarely investigated in the available literature. To this end, this paper aims to present the flow distribution and resistance characteristics of the PCHE plate, which was studied with numerical models under different flow distribution cases; the numerical simulation of fluid flow distribution can provide guidance for subsequent research and the design and development of multi-channel heat exchangers.

2. Physical Problems and Mathematical Model

2.1. Physical Problems

Due to the limitations of the processing technology and conditions, it is difficult to process large-sized PCHEs. Presently, the use of square plates is an effective way to improve the heat transfer power of a single heat exchanger unit. However, the inlet and outlet parts occupy a large heat transfer area in the nearly square plate using the traditional single-in and single-out flow, resulting in the area of the counter-current heat transfer part in the middle being very limited, as shown in Figure 1, and greatly influencing the heat transfer efficiency of PCHE. The use of a collecting flow in the inlet and outlet will greatly reduce the space occupied by the inlet and outlet parts which, in turn, will maximize the length of the counter-current heat transfer part, as shown in Figure 2. The flow distribution was very inhomogeneous on the heat exchanger plate between the inlet and outlet, due to the different pressures of the different channels, which will lead to a significant decrease in the heat transfer efficiency of PCHE and an increase in the flow resistance [19].

Therefore, the uniformity of flow distribution is a key factor affecting the performance of PCHE [20]. In this paper, a PCHE numerical model was established. The flow distribution and resistance characteristics of a PCHE with collecting flow plates at the inlet and outlet were studied by using different flow distribution cases.

The calculation model is shown in Figure 3, combining the existing problems, and the heat exchanger plate was set with a collecting flow at the inlet and outlet. Some spreader plates were placed in the inlet and outlet parts to improve the uniformity of flow distribution within the channels. The width between the inlet and outlet is B, the width between spreader plates is b, and the radius of spreader plates are r_1, r_2, r_3, r_4, r_5 respectively.

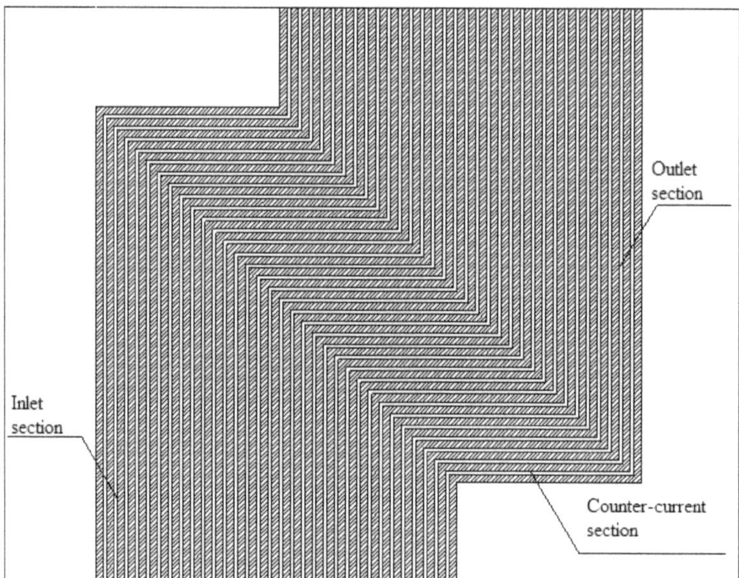

Figure 1. Heat exchanger plate with single inlet and outlet.

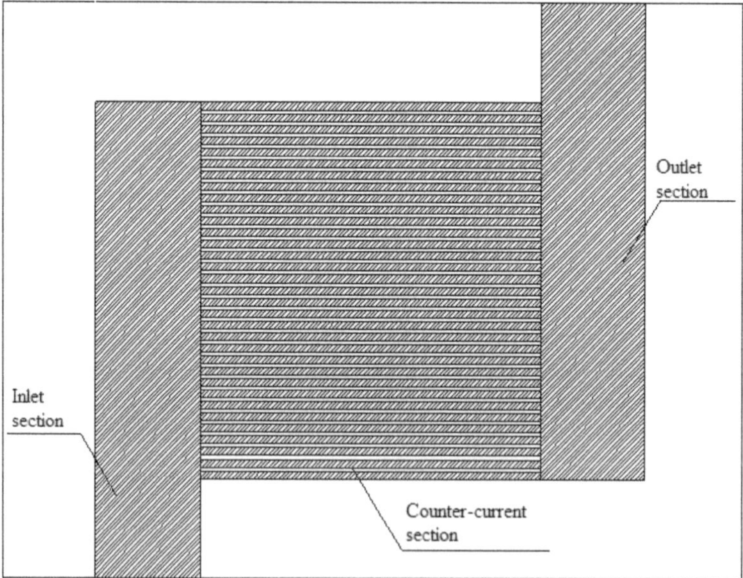

Figure 2. Heat exchanger plate with collecting flow at the inlet and outlet.

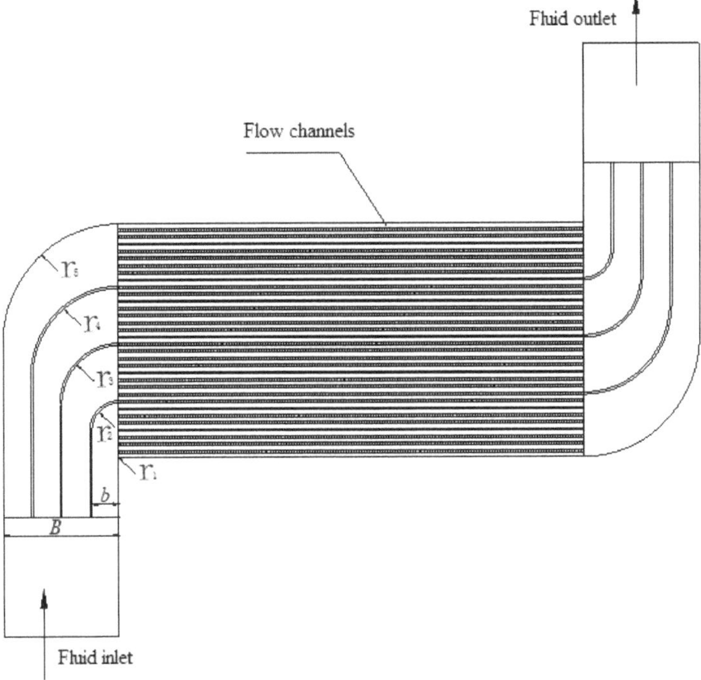

Figure 3. Calculation model for the multi-channel heat exchanger.

2.2. Assumptions

In order to facilitate the study of the computational model, the following assumptions must be defined:

(1) Flow is steady and isothermal, and fluid properties are independent of time;
(2) Fluid density is dependent on the local temperature only, or is treated as a constant;
(3) Fluid slip at the solid-fluid interfaces is neglected;
(4) Thermo-physical properties of fluid are independent of temperature variations;
(5) Body forces are caused only by gravity (i.e., magnetic, electrical, and other fields do not contribute to the body forces);
(6) Newtonian fluid, incompressible flow and turbulent flow.

2.3. Governing Equations

The governing equations describe the operation of fluid flow inside the channel. When the governing equations were used to directly model the flow, it was necessary to use small time and space steps to distinguish the detailed spatial structure and the time-varying temporal characteristics in turbulence. However, this requires a large amount of memory space and a very high CPU operating speed. Therefore, the direct numerical calculation of the governing equation is still difficult to use in engineering calculations. However, if the turbulence models in FLUENT (like the k-ε model) are introduced to simplify the N-S equation, it can be applied to engineering calculations. The general form of all the governing equations in the fluid domain for the present steady-state problem is shown as follows:

$$\mathrm{div}\left(\rho\vec{V}\phi\right) = \mathrm{div}[\Gamma \cdot grad(\phi)] + S \tag{1}$$

It should be noted that the above equation can be used for describing the equations of continuity, momentum, turbulent dissipation rate and turbulence energy.

For the equation of continuity:

$$\phi = 1,\ \Gamma = 0,\ S = 0 \tag{2}$$

For the equations of momentum (u, v, w):

$$u:\ \phi = u,\ \Gamma = \eta_{eff},\ S = -\frac{\partial p}{\partial x} + \eta_{eff}\left(\frac{\partial^2 u}{\partial x^2} + \frac{\partial^2 v}{\partial x \partial y} + \frac{\partial^2 w}{\partial x \partial z}\right) \tag{3a}$$

$$v:\ \phi = v,\ \Gamma = \eta_{eff},\ S = -\frac{\partial p}{\partial y} + \eta_{eff}\left(\frac{\partial^2 u}{\partial y \partial x} + \frac{\partial^2 v}{\partial y^2} + \frac{\partial^2 w}{\partial y \partial z}\right) \tag{3b}$$

$$w:\ \phi = w,\ \Gamma = \eta_{eff},\ S = -\frac{\partial p}{\partial z} + \eta_{eff}\left(\frac{\partial^2 u}{\partial z \partial x} + \frac{\partial^2 v}{\partial z \partial y} + \frac{\partial^2 w}{\partial z^2}\right) \tag{3c}$$

For the equation of turbulent dissipation rate (k):

$$\rho u_j \frac{\partial k}{\partial x_j} = \frac{\partial}{\partial x_j}\left[\left(\eta + \frac{\eta_t}{\sigma_k}\right)\frac{\partial k}{\partial x_j}\right] + \eta_t \frac{\partial u_i}{\partial x_j}\left(\frac{\partial u_i}{\partial x_j} + \frac{\partial u_j}{\partial x_i}\right) - \rho\varepsilon \tag{4}$$

For the equation of turbulence energy (k):

$$\rho u_k \frac{\partial \varepsilon}{\partial x_k} = \frac{\partial}{\partial x_k}\left[\left(\eta + \frac{\eta_t}{\sigma_\varepsilon}\right)\frac{\partial \varepsilon}{\partial x_k}\right] + \frac{c_1 \varepsilon}{k}\eta_t \frac{\partial u_i}{\partial x_j}\left(\frac{\partial u_i}{\partial x_j} + \frac{\partial u_j}{\partial x_i}\right) - c_2 \rho \frac{\varepsilon^2}{k} \tag{5}$$

η_t can be written as $\eta_t = c'_\mu \rho k^{\frac{1}{2}} l = \left(c'_\mu c_D\right)\rho k^2 \frac{1}{c_D k^{\frac{3}{2}}/l} = c_\mu \rho k^2/\varepsilon$. In the above equations, the parameter ρ means the density of the fluid, S represents the source item; l means the characteristic length of the turbulent channel; k means the turbulence energy; ε means the turbulent dissipation rate; C_D, C_1, C_2 means the empirical constant.

2.4. Boundary Conditions

Inlet: Velocity Inlet Boundary Condition

In the velocity inlet boundary condition, the stagnation point parameters of the inlet boundary in the flow field are not fixed. In order to satisfy the velocity condition at the inlet, the stagnation point parameter will fluctuate within a certain range.

Outlet: Outflow Outlet Boundary Condition

The boundary condition of free outflow is subject to the assumption of fully developed turbulence. Fully developed flow means that the flow field variables do not change in the flow direction, that is, the diffusion flux of all flow variables is equal to zero in the normal direction of the outlet boundary.

2.5. Numerical Details

The viscosity in the solid domain is considered infinite. The thermo-physical properties of the working fluid can be calculated based on its temperature and pressure, as shown in Table 1. In this table, some of the permanent parameters in the numerical simulation are clearly presented, including channel type, number of channels, plate size, channel size, channel spacing, the width of the inlet and the outlet, the working fluid, pressure, and inlet velocity.

Table 1. PCHE parameters.

Parameter	Value
Channel type	Square straight channel
Number of channels	33
Plate size	300 mm × 200 mm
Channel size	2 mm × 1 mm
Channel spacing	1 mm
Inlet and outlet width	50 mm
Working fluid	water
Pressure	0.3 MPa
Inlet velocity	0–40 m/s

3. Results and Discussion

3.1. Grid Independence Check

For the grid-independence verification, five numerical calculation steps with 1, 0.5, 0.3, 0.2 and 0.1 mm were carried out in this paper. As shown in Section 2, the numerical model had 33 channels in the counter-current section. Figure 4 shows the flow rate variations of two representative channels (1 and 33). It can be seen that, when the grid number reached 80,850 with a step size of 0.2 mm, the results changed a little (<1%) with the increase in the grid number, which means a step size of 0.2mm was small enough for this numerical model. Therefore, a step size of 0.2 mm was used in the subsequent calculations.

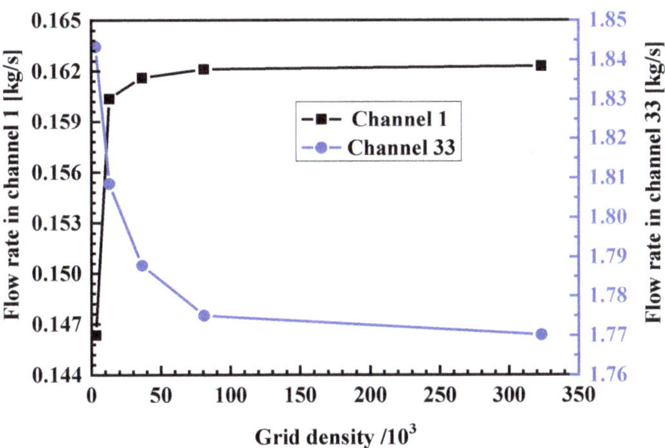

Figure 4. Validation of grid independence.

3.2. Analysis of Flow Maldistribution

In order to evaluate the flow distribution characteristics within the channels, 33 channels were numbered from bottom to top along the flow direction from the inlet. In order to compare the flow distribution uniformity of 33 channels under different flow rates, two parameters, i.e., flow distribution nonuniformity in each PCHE channel E_i and the average flow distribution nonuniformity in all PCHE channels S are defined as follows:

$$E_i = \frac{m_i - m_a}{\sum_{i=1}^{n} m_i} \tag{6}$$

$$S = \sqrt{\frac{1}{n-1}\sum_{i=1}^{n}\left(\frac{m_i}{m_a} - 1\right)^2} \tag{7}$$

where, E_i is used to characterize the flow distribution nonuniformity between different channels, S is used to characterize the total flow distribution nonuniformity of the heat exchanger, m_i is the flow rate in the ith channel, m_a is the average flow rate for one single channel and n is the number of channels.

3.3. Flow Distribution Case

In order to improve the uniformity of traffic distribution in the channel, this paper studied the flow distribution characteristics in different cases, as shown in Figure 5. The parameters of the different cases are shown in Table 2. Case 1 used no spreader plate. Case 2 used the spreader plate with an increase in the inner and outer radius. Case 3 used the spreader plate with an equal inner and outer radius. Case 4 used the spreader plate with an equal inner and outer radius, also with a large bending radius.

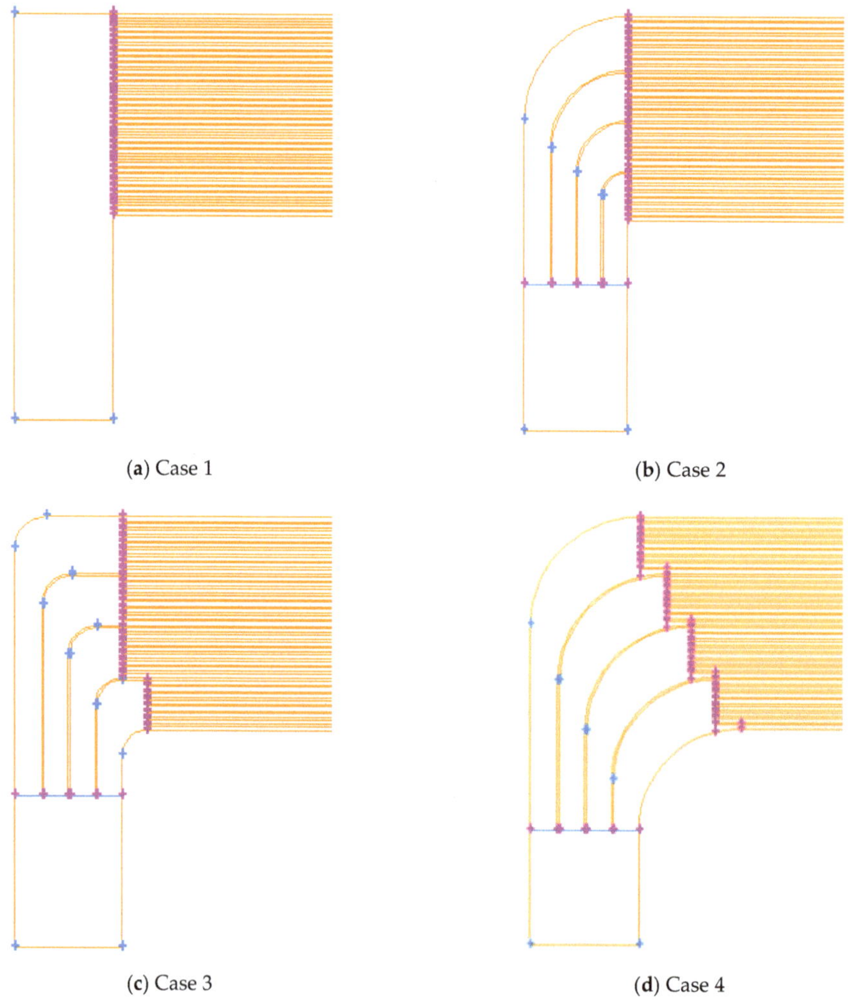

(**a**) Case 1 (**b**) Case 2

(**c**) Case 3 (**d**) Case 4

Figure 5. Configurations of different cases.

Table 2. Parameters of PCHE plates for flow distribution.

Cases	Parameters
1	No spreader plate
2	$r_1 = 0, r_2 = b, r_3 = 2b, r_4 = 3b, r_5 = 4b$
3	$r_1 = r_2 = r_3 = r_4 = r_5 = b$
4	$r_1 = r_2 = r_3 = r_4 = r_5 = 4b$
Inlet velocity	0–40 m/s

Figure 6 shows the streamlines for the different cases with an inlet velocity of 40 m/s. It was observed that, when the spreader plate is not used (Case 1), the angle of the fluid was large when fluid flowed into the channel, which caused a clear flow separation and a strong eddy current at the inlet of the channel. The flow separation area occupied the largest section. When the spreader plates in Cases 2 and 3 were used, the flow separation at the inlet of the channel gradually improved, and the flow separation and vortex area were greatly reduced. When the spreader plate in Cases 4 was used, the radius of the spreader plate was greatly increased, but the angle of the fluid was greatly reduced, which meant there was basically no flow separation in the channel.

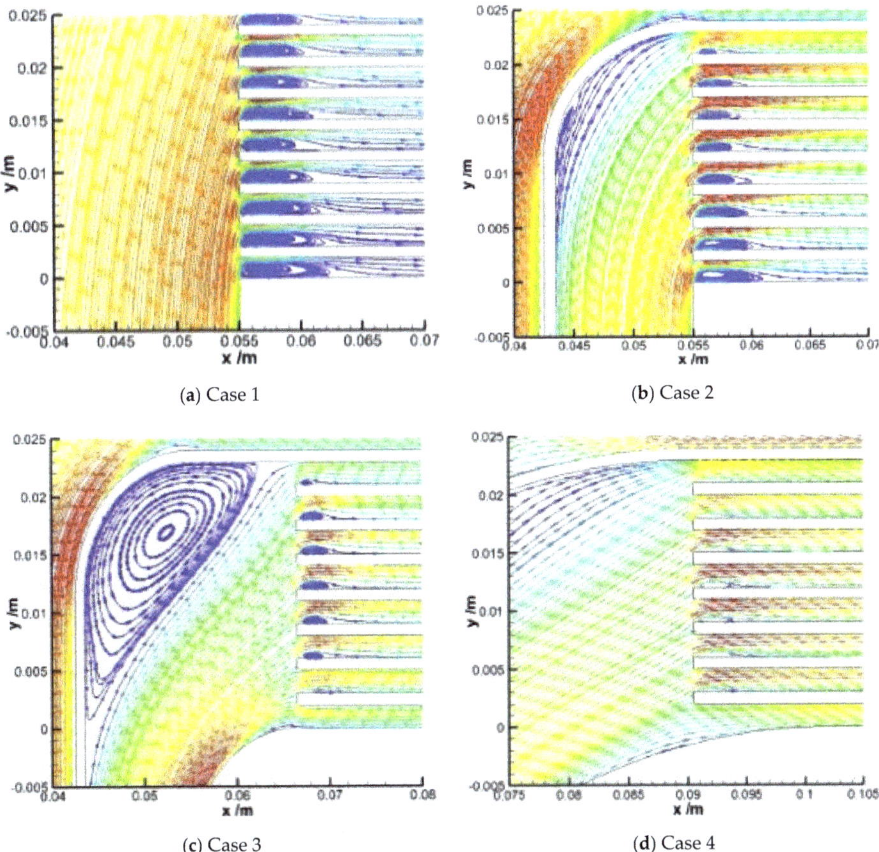

(**a**) Case 1 (**b**) Case 2

(**c**) Case 3 (**d**) Case 4

Figure 6. Streamlines of different cases.

Figure 7 shows the flow distribution in the channel for the different cases with an inlet velocity of 40 m/s. Compared with Figure 6, it can be seen the flow distribution in different channel was very inhomogeneous due to the violent flow separation at the inlet of the channel in Case 1. The flow at the inlet of the channel in Case 2 and 3 improved gradually, but the flow separation and vortex area greatly decreased, and the imbalance of flow distribution of the channel greatly improved. There was barely any flow separation phenomenon at the inlet of the channel in Case 4, and the imbalance of the flow distribution was also decreased with ±0.8%.

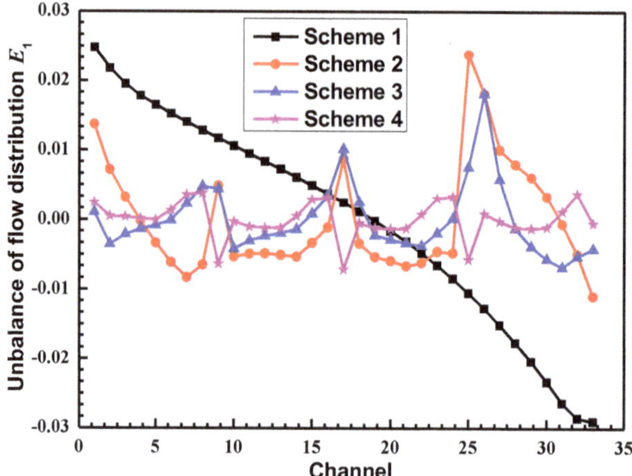

Figure 7. Imbalance of flow distribution for different cases.

Figure 8 shows the nonuniformity variation in the total flow distribution with the different inlet velocity. It was observed that the nonuniformity of flow distribution decreased from Case 1 to Case 4, which indicates that the uniform flow distribution corresponded with the smaller flow rate and dispersion of the average flow in the channel. With an increase in the inlet velocity, the nonuniformity of the flow distribution of each case increased. In the inlet velocity range, as shown in Figure 8, the nonuniformity of the flow distribution was within 9%, indicating that the uniformity of the flow distribution was good.

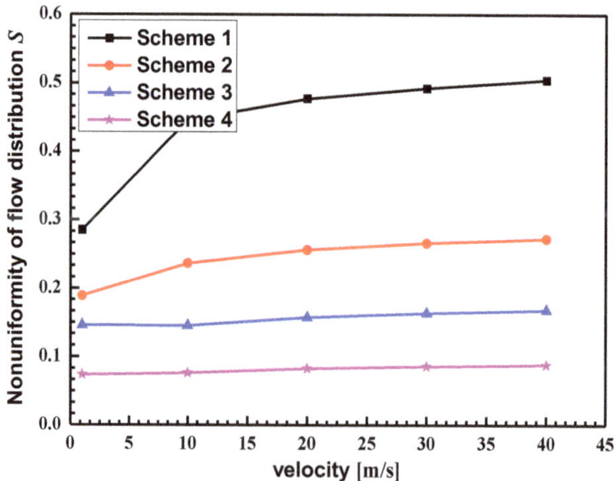

Figure 8. Nonuniformity of flow distribution for different cases.

Figure 9 shows the variation in pressure drop between the inlet and outlet with different velocities. Compared with Figures 6, 8 and 9, the pressure drop was great in Case 1 due to a very serious flow separation, which caused an inhomogeneous flow distribution. When the spreader plate was added in Cases 2 and 3, the flow separation improved and the flow distribution was more uniform, so the flow resistance was greatly reduced. There was barely any flow separation phenomenon in Case 4. The flow distribution uniformity in this case was the best. Therefore, the flow resistance was also the minimum.

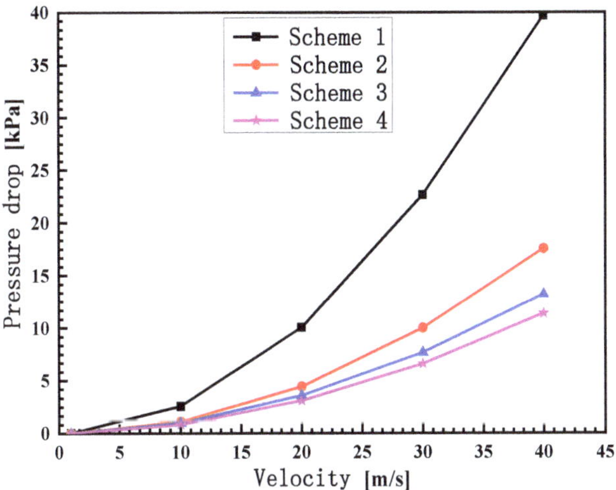

Figure 9. Variation in the pressure drop with different velocities.

3.4. Numbers of Spreader Plate

In order to improve the uniformity of the flow distribution in the channel further, the flow distribution characteristics of different spreader plates were studied based on Case 4, as shown in Figure 10. The radius of the spreader plate was set as $r_1 = r_2 = r_3 = r_4 = r_5 = 4b$.

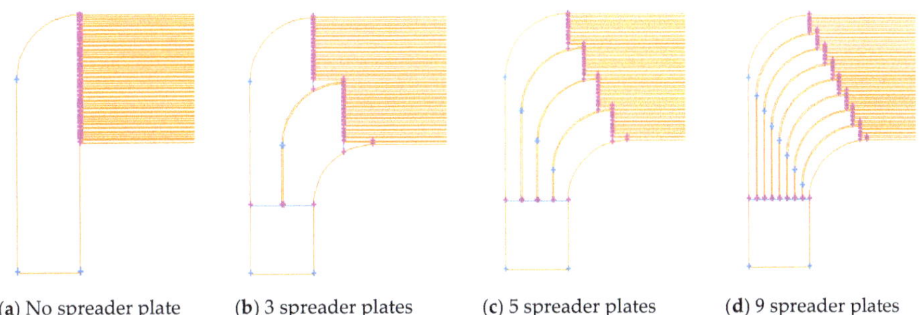

(**a**) No spreader plate (**b**) 3 spreader plates (**c**) 5 spreader plates (**d**) 9 spreader plates

Figure 10. Different numbers of spreader plates.

Figure 11 shows the flow distribution in the channel with different numbers of spreader plates. It was shown that with an increase in the spreader plates and partitions, the imbalance of the flow distribution in the channel was greatly improved. When the number of spreader plates reached nine, the imbalance of the flow distribution was reduced to ±0.2%.

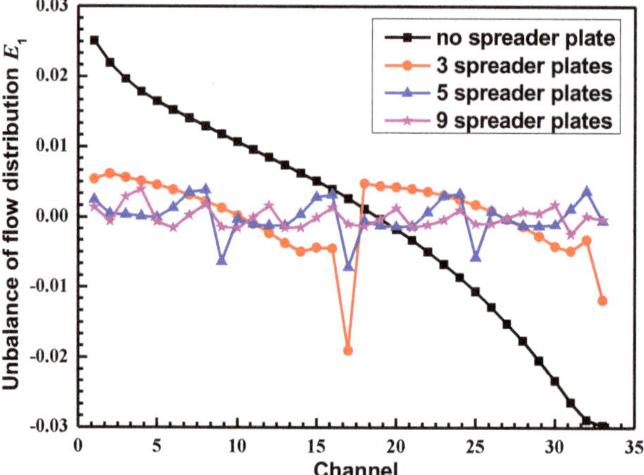

Figure 11. Imbalance of flow distribution for different numbers of spreader plates.

Figure 12 shows the nonuniformity variation in the total flow distribution with the inlet velocity under different spreader plates. With an increase in the number of spreader plates, the nonuniformity of the flow distribution decreased, indicating that the uniform flow distribution corresponded with the smaller flow rate and dispersion of the average flow in the channel. With an increase in the inlet flow rate, the nonuniformity of the flow distribution of each model increased. The nonuniformity of the flow distribution with no spreader plates and three spreader plates varied greatly with the velocity, but with five and nine spreader plates it only varied slightly. When the number of spreader plates reached nine, the flow distribution in the channel was less than 5%, indicating that the flow distribution was very uniform.

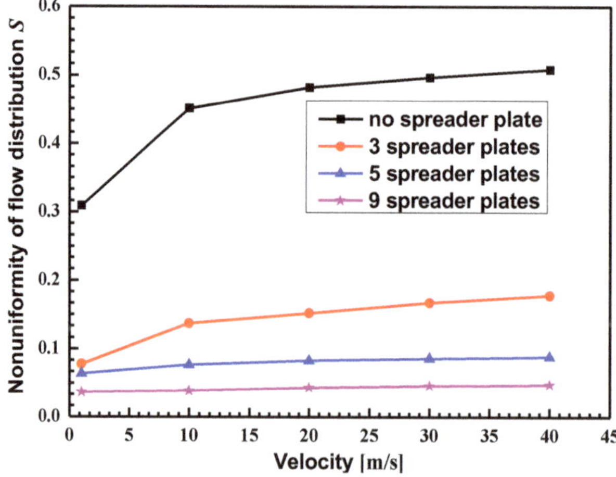

Figure 12. Nonuniformity of flow distribution for different numbers of spreader plates.

Figure 13 shows the variation in pressure drop between the inlet and outlet with velocities under different spreader plates. Compared with Figures 9 and 13, the pressure drop was large when there were no spreader plates. When the spreader plates were added, the flow resistance was largely decreased. It was also observed that the pressure loss of nine spreader plates was larger than that of three and five spreader plates. It can be explained that, when the number of spreader plates increased, the flow cross-sectional area at the inlet and outlet increased, the flow velocity increased, and the pressure loss increased. This indicated that the number of spreader plates could not increase without limitation, which should comply with the integrated uniformity of the flow distribution, flow resistance and heat transfer efficiency.

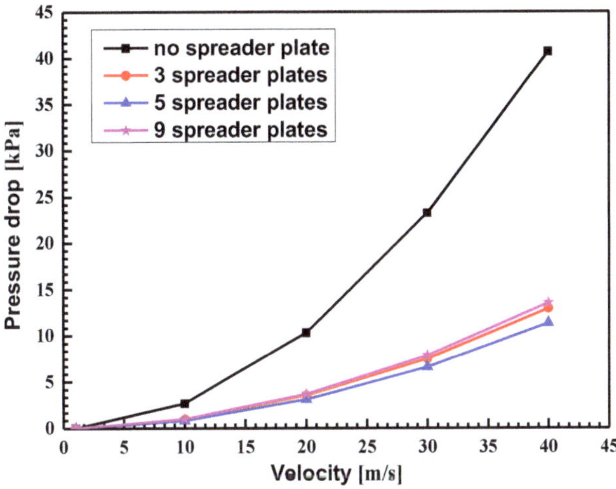

Figure 13. Variation in the pressure drop with different numbers of spreader plates.

3.5. *Analysis of the Symmetry*

3.5.1. Structure Central Symmetry

The whole structure of PCHE shows central symmetry. The flow distribution and resistance characteristics of a PCHE plate were studied with numerical models under different flow distribution cases. Although only the inlet part of the heat exchanger was considered and modeled in this paper, the conclusions of this paper are also applicable to the outlet part of the heat exchanger. The results show that the sudden change in the angle of the fluid at the inlet of the channel can be greatly reduced by using the spreader plate with equal inner and outer radius.

3.5.2. Channel Flow Field Symmetry

The streamlines of the flow field in the PCHE channel presents symmetry. As Figure 6 shows, there was barely any flow separation phenomenon at the inlet of the channel in Case 4 (the spreader plate with an equal inner and outer radius, also with a large bending radius), and the imbalance of the flow distribution was also decreased by ±0.8%. For the flow field at the outlet of the heat exchanger, the streamlines of the flow field exhibited the same characteristics as the inlet.

3.5.3. Adjacent Channel Structure Symmetry

The geometry of each channel of the PCHE presents structural symmetry. Therefore, the velocity field and temperature field of each channel also exhibit certain symmetric characteristics. Improving the symmetry of the adjacent channels can diminish the flow rates in adjacent channels. One goal of this study was to improve the symmetry of the structures of adjacent channels and thus ensure the symmetry of the fluid flow in adjacent channels, in order to diminish the flow rates in adjacent channels and, especially, to prevent maldistribution in the channels near the side walls of the device.

3.6. Discussion

This paper aims to present the flow distribution and resistance characteristics of PCHE plates, which were studied with numerical models under different flow distribution cases; the numerical simulation of fluid flow distribution can provide guidance for subsequent research and the design and development of multi-channel heat exchangers.

The multi-channel PCHE plate showed an obvious maldistribution effect in the practical application, which produced a great limitation in the performance promotion of the PCHE plate. This aspect was rarely investigated in the available literature. This paper analyzes the symmetry of the heat exchanger in three aspects: structure central symmetry, channel flow field symmetry and adjacent channel structure symmetry.

The results show that the sudden change in the angle of the fluid at the inlet of the channel can be greatly reduced by using the spreader plate with an equal inner and outer radius. The flow separation of the fluid at the inlet of the channel can also be weakened and the imbalance of flow distribution in the channel can be reduced. The numerical simulation of the fluid flow distribution can provide guidance for the subsequent research and the design and development of multi-channel heat exchangers.

4. Conclusions

In this study, different flow distribution cases were designed and theoretically studied in order to improve the uniformity of the fluid flow distribution, and the resistance characteristics at the inlet and outlet of the PCHE channel were numerically modeled and analyzed in detail. Our conclusions can be drawn as follows:

(1) With the use of the spreader plate with an equal inner and outer radius, the sudden change in the angle of the fluid at the inlet of the channel can be greatly reduced. The flow separation of the fluid at the inlet of the channel can also be weakened and the imbalance of flow distribution in the channel can be reduced. So, the flow uniformity can be improved and the pressure loss between the inlet and outlet of the PCHE can be reduced;

(2) With the use of nine spreader plates, the imbalance of flow distribution in the channel is basically reduced to $\pm 0.2\%$, and the total flow distribution is within 5%, which indicates a good flow distribution uniformity;

(3) The uniformity of the flow distribution increases with the increase in spreader plates. However, with the increase in the number of spreader plates, the flow cross-sectional area at the inlet and outlet increases, the flow velocity increases, and the pressure loss increases. Therefore, the number of spreader plates cannot increase without limitation, which should comply with the integrated uniformity of the flow distribution, flow resistance and heat transfer efficiency.

Author Contributions: H.K. and H.X. led the writing of the paper. H.K., Y.L., Z.K., Q.X., Z.W., K.C. and H.X. developed the numerical model. H.K., Y.L., Z.K., Q.X., Z.W., K.C. and H.X. performed the data analysis. The first version was drafted by H.K. and H.X. All authors have read and agreed to the published version of the manuscript.

Funding: This research was funded by National Natural Science Foundation of China (Nos. 51876146, 51876118), Open Fund of Science and Technology on Thermal Energy and Power Laboratory (No. TPL2018B03), and Shanghai International Science and Technology Cooperation Fund (No. 18160743800).

Acknowledgments: Chenqian Wu in Shanghai Jiao Tong University is acknowledged for some helpful discussion and useful information in improving the quality of this paper.

Conflicts of Interest: There is no conflict of interest.

References

1. Shah, R.K.; Sekulic, D.P. *Fundamentals of Heat Exchanger Design*; John Wiley & Sons: Hoboken, NJ, USA, 2003.
2. Ezgi, C.; Özbalta, N.; Girgin, I. Thermohydraulic and thermoeconomic performance of a marine heat exchanger on a naval surface ship. *Appl. Therm. Eng.* **2014**, *64*, 413–421. [CrossRef]

3. Kakaç, S.; Liu, H.; Pramuanjaroenkij, A. *Heat Exchangers: Selection, Rating, and Thermal Design*; CRC Press: New York, NY, USA, 2002.
4. Farajollahi, B.; Etemad, S.G.; Hojjat, M. Heat transfer of nanofluids in a shell and tube heat exchanger. *Int. J. Heat Mass Transf.* **2010**, *53*, 12–17. [CrossRef]
5. Tsuzuki, N.; Kato, Y.; Ishiduka, T. High performance printed circuit heat exchanger. *Appl. Therm. Eng.* **2007**, *27*, 1702–1707. [CrossRef]
6. Sabharwall, P.; Kim, E.S.; McKellar, M.; Anderson, N. *Process Heat Exchanger Options for the Advanced High Temperature Reactor*; Idaho National Laboratory (INL): Idaho Falls, 2011.
7. Bartel, N.; Chen, M.; Utgikar, V.; Sun, X.; Kim, I.-H.; Christensen, R.; Sabharwall, P. Comparative analysis of compact heat exchangers for application as the intermediate heat exchanger for advanced nuclear reactors. *Ann. Nucl. Energy.* **2015**, *81*, 143–149. [CrossRef]
8. Shin, C.W.; No, H.C. Experimental study for pressure drop and flow instability of two-phase flow in the PCHE-type steam generator for SMRs. *Nucl. Eng. Des.* **2017**, *318*, 109–118. [CrossRef]
9. Ma, T.; Li, L.; Xu, X.-Y.; Chen, Y.-T.; Wang, Q.-W. Study on local thermal–hydraulic performance and optimization of zigzag-type printed circuit heat exchanger at high temperature. *Energy Convers. Manag.* **2015**, *104*, 55–66. [CrossRef]
10. Yang, H.; Wen, J.; Gu, X.; Liu, Y.; Wang, S.; Cai, W.; Li, Y. A mathematical model for flow maldistribution study in a parallel plate-fin heat exchanger. *Appl. Therm. Eng.* **2017**, *121*, 462–472. [CrossRef]
11. Jiao, A.; Zhang, R.; Jeong, S. Experimental investigation of header configuration on flow maldistribution in plate-fin heat exchanger. *Appl. Therm. Eng.* **2003**, *23*, 1235–1246. [CrossRef]
12. Bobbili, P.R.; Sunden, B.; Das, S.K. An experimental investigation of the port flow maldistribution in small and large plate package heat exchangers. *Appl. Therm. Eng.* **2006**, *26*, 1919–1926. [CrossRef]
13. Zhang, Z.; Mehendale, S.; Tian, J.; Li, Y. Experimental investigation of distributor configuration on flow maldistribution in plate-fin heat exchangers. *Appl. Therm. Eng.* **2015**, *85*, 111–123. [CrossRef]
14. Zhang, Z.; Li, Y. CFD simulation on inlet configuration of plate-fin heat exchangers. *Cryogenics* **2003**, *43*, 673–678. [CrossRef]
15. Wen, J.; Li, Y.; Zhou, A.; Zhang, K. An experimental and numerical investigation of flow patterns in the entrance of plate-fin heat exchanger. *Int. J. Heat Mass Transf.* **2006**, *49*, 1667–1678. [CrossRef]
16. Sheik Ismail, L.; Ranganayakulu, C.; Shah, R.K. Numerical study of flow patterns of compact plate-fin heat exchangers and generation of design data for offset and wavy fins. *Int. J. Heat Mass Transf.* **2009**, *52*, 3972–3983. [CrossRef]
17. Wei, Z.; Yang, W.; Xiao, R. Pressure Fluctuation and Flow Characteristics in a Two-Stage Double-Suction Centrifugal Pump. *Symmetry* **2019**, *11*, 65. [CrossRef]
18. Afridi, M.I.; Tlili, I.; Goodarzi, M.; Osman, M.; Khan, N.A. Irreversibility analysis of hybrid nanofluid flow over a thin needle with effects of energy dissipation. *Symmetry* **2019**, *11*, 663. [CrossRef]
19. Shi, H.-N.; Ma, T.; Chu, W.-X.; Wang, Q.-W. Optimization of inlet part of a microchannel ceramic heat exchanger using surrogate model coupled with genetic algorithm. *Energy Convers. Manag.* **2017**, *149*, 988–996. [CrossRef]
20. Koo, G.-W.; Lee, S.-M.; Kim, K.-Y. Shape optimization of inlet part of a printed circuit heat exchanger using surrogate modeling. *Appl. Therm. Eng.* **2014**, *72*, 90–96. [CrossRef]

Article

Mathematical Analysis on an Asymmetrical Wavy Motion of Blood under the Influence Entropy Generation with Convective Boundary Conditions

Arshad Riaz [1], Muhammad Mubashir Bhatti [2], Rahmat Ellahi [3,4,*], Ahmed Zeeshan [3] and Sadiq M. Sait [5]

1 Department of Mathematics, Division of Science and Technology, University of Education, Lahore 54770, Pakistan; arshad-riaz@ue.edu.pk
2 College of Mathematics and Systems Science, Shandong University of Science and Technology, Qingdao 266590, China; mmbhatti@sdust.edu.cn
3 Department of Mathematics & Statistics, Faculty of Basic and Applied Sciences (FBAS), International Islamic University (IIUI), Islamabad 44000, Pakistan; ahmad.zeeshan@iiu.edu.pk
4 Fulbright Fellow Department of Mechanical Engineering, University of California Riverside, Riverside, CA 92521, USA
5 Center for Communications and IT Research, Research Institute, King Fahd University of Petroleum & Minerals, Dhahran 31261, Saudi Arabia; sadiq@kfupm.edu.sa
* Correspondence: rellahi@alumni.ucr.edu or rahmatellahi@yahoo.com

Received: 13 December 2019; Accepted: 3 January 2020; Published: 6 January 2020

Abstract: In this article, we discuss the entropy generation on the asymmetric peristaltic propulsion of non-Newtonian fluid with convective boundary conditions. The Williamson fluid model is considered for the analysis of flow properties. The current fluid model has the ability to reveal Newtonian and non-Newtonian behavior. The present model is formulated via momentum, entropy, and energy equations, under the approximation of small Reynolds number and long wavelength of the peristaltic wave. A regular perturbation scheme is employed to obtain the series solutions up to third-order approximation. All the leading parameters are discussed with the help of graphs for entropy and temperature profiles. The irreversibility process is also discussed with the help of Bejan number. Streamlines are plotted to examine the trapping phenomena. Results obtained provide an excellent benchmark for further study on the entropy production with mass transfer and peristaltic pumping mechanism.

Keywords: convection; entropy production; heat transfer engineering; blood flow

1. Introduction

In our daily life, living organisms require energy to do physical work and keep the body temperature under the influence of heat exchange to the environment, as well as to generate, replace, and propagate molecules to the relevant constituents. Such type of energy comes from the oxidation process of organic substances i.e., amino acids, fats, and carbohydrates fed to the organisms. As compared to the other heat engines (i.e., in which the chemical energy gets transformed to the thermal energy, and then is transformed to mechanical work), living organisms can transform the nutrient's chemical energy into work. It happens due to the oxidation of nutrients located internally in the organisms i.e., metabolism, pass through different steps, which helps to hold some energy from ATP (adenosine triphosphate). The ATP utilized entirely by all beings for the direct transformation of mechanical energy and also actively supports other biological reactions [1]. In recent years, various authors [2–6] have examined the heat production of mammals via calorimetry, and presented that for the given nutrients, both combustion and animal metabolism expends the same amount of oxygen.

According to previous research [7], it is found that living things can produce thermal energy via fat, and combustion of carbohydrates in the living body, and is identical to the oxidation of heat of these elements. As a result, the amount of nutrients digested by a living being, and hence its input energy, can be determined by the chemical composition of food intake and the measurements of breathing i.e., CO_2 and O_2.

Hershey [8] and Hershey and Wang [9] examined the entropy production during the lifespan of a human being. They found that when the human body is in a state of rest, mostly the output energy due to the nutrient's metabolism occurs as a heat. They also reinforced the calorimeter to examine the heat transfer rate to the environment and verified that with the help of BMR (basal metabolic). According to their results, it was found that entropy generated over a lifespan was 10,678 KJ/kg·K for females, and 10,025 KJ/kg·K for males. Rahman [10] discussed the entropy generation for forced and free convection using a new mathematical model. He discussed forced and free convection at distinct mass influx, outflux (i.e., waste, air, food, and water, etc.), level of physical activity, clothing effects, and airspeeds. His results are similar to those from Hershey [8] and Aoki [11] but one order of magnitude higher in general. Annamalai and Puri [12] utilized the first law of thermodynamics to achieve the metabolic scaling for a biological system. They also used the second law of thermodynamics to determine the entropy generation in humans and prognosticate the lifespan of 77 years by assuming the maximal entropy generation as 10,000 KJ/kg·K. Bejan [13,14] introduced a constructal design principle and presented optimal geometric types scales to the power of their associated size and showed that different natural structures (i.e., lightening, river deltas, tree branches, and vascularized tissue) are periodic in nature. Rashidi et al. [15] discussed the entropy generation with magnetic effects and slip boundary conditions propagating among an infinite porous disk having variable features. According to their results, they observed that the disk is an essential root of entropy generation. Komurgoz et al. [16] examined the entropy generation through an inclined porous channel with magnetic effects. According to their results, they found that maximal entropy production can be gained in the absence of porosity and magnetic field.

Blood is an essential part of the human body, which comprises 7% of the total body weight. The leading role of blood is molecular oxygen for cellular metabolism and carry the nutrients as well as a significant role in thermoregulatory. Blood performs as a non-Newtonian fluid. The blood viscosity changes due to the shear rate. The viscosity of the blood can be analyzed by the hematocrit, plasma (constitute 54.3% of the whole blood) viscosity, and the mechanical features of red blood cells (constitute 45% of the whole blood). The human blood is a heterogeneous solution that contain multiple kinds of cells (known as corpuscles or formed elements), which consist of leukocytes, thrombocytes, and erythrocytes. In view of such importance, different authors discussed the entropy generation in blood. For instance, Rashidi et al. [17] discussed the magnetic effects on the blood flow propagating through a porous medium with a filtration and control process. Akbar et al. [18] examined the thermal conductivity on the peristaltic propulsion of H_2+Cu nanofluids with entropy production. Rashidi et al. [19] obtained the series solution for the entropy generation of the blood flow of a nanofluid in the presence of a magnetic field. Endoscopic effect and entropy production on peristaltic nanofluid flow, having a thermal conductivity of 2 HO were investigated by Akbar et al. [20]. Abbas et al. [21] presented a detailed analysis of the peristaltic flow with nanofluids and entropy production through a finite channel with compliant walls. Bhatti et al. [22] considered the Casson blood flow to examine the entropy process with peristaltic movement under the uniform magnetic field. Ranjit and Shit [23] examined the entropy production on the electroosmotic flow under uniform magnetic field with peristaltic pumping. More studies on the blood flow and entropy generation can be found from the references [24–28].

According to the above survey, it is found that less attention has been given to the entropy production asymmetric peristaltic propulsion of blood flow with heat transfer. Therefore, in the present analysis, we discuss the entropy generation with convection on the asymmetric propulsion of the peristaltic blood of nonlinear Williamson fluid. An assumption of long peristaltic wavelength is taken into account and Reynolds number is considered to be very small (Re $\approx$ 0). A regular perturbation method is used to obtain series solutions. The novelty of all the leading parameters is discussed and illustrated. The trapping mechanism is also examined to determine the nonlinear asymmetric peristaltic motion.

2. Governing Equations

In this section we analyze the incompressible peristaltic propulsion of Williamson fluid in a two-dimensional channel with a width $\overline{d_1} + \overline{d_2}$. The flow is initialized by a sinusoidal wave propagating with a constant speed c along the layout of channel (see Figure 1). The addition here is the extra equations of energy and entropy generation. It is assumed that the temperature at the upper wall of the channel is T_1 and lower wall has the temperature T_0 such that $T_0 < T_1$. It depicts the physical reasoning that heat will transfer from lower to upper wall. The wall surfaces are suggested as:

$$\overline{Y} = \begin{cases} H_i = \overline{d_i} + \overline{a_i} \cos\left[2\pi\omega\right], \ i = 1, \\ H_j = -\overline{d_j} - \overline{b_i} \cos\left[2\pi\omega + \phi\right], \ j = 2, \end{cases} \tag{1}$$

and

$$\omega = \frac{\overline{X} - c\overline{t}}{\lambda}, \tag{2}$$

where $\overline{a_1}$ and $\overline{b_1}$ are the wave amplitudes, λ the wave length, $\overline{t}$ the time, c the velocity of the propagation, and $\overline{X}$ is the direction of wave propagation. The phase difference ϕ has the range $0 \leq \phi \leq \pi$ i.e., waves out of phase $\phi = 0$ associated to the symmetric channel, and $\phi = \pi$ associated to the waves are in phase. Moreover, $\overline{a_i}$, $\overline{b_i}$, $\overline{d_i}$, $\overline{d_j}$, and ϕ satisfy the condition:

$$\overline{a_i}^2 + \overline{b_i}^2 + 2\overline{a_i}\overline{b_i} \cos\phi \leq \left(\overline{d_i} + \overline{d_j}\right)^2. \tag{3}$$

The equations of momentum in component forms are described as:

$$\rho\left[D_{\overline{t}} + \overline{U} D_{\overline{X}} + \overline{V} D_{\overline{Y}}\right] \overline{U} = -D_{\overline{X}} P + D_{\overline{X}} \overline{S}_{\overline{XX}} + D_{\overline{Y}} \overline{S}_{\overline{XY}}, \tag{4}$$

where $D_{\overline{t}} = \frac{\partial}{\partial \overline{t}}, D_{\overline{X}} = \frac{\partial}{\partial \overline{X}}, D_{\overline{Y}} = \frac{\partial}{\partial \overline{Y}}$.

$$\rho\left[D_{\overline{t}} + \overline{U} D_{\overline{X}} + \overline{V} D_{\overline{Y}}\right] \overline{V} = D_{\overline{Y}} P + D_{\overline{X}} \overline{S}_{\overline{YX}} + D_{\overline{Y}} \overline{S}_{\overline{YY}}. \tag{5}$$

The stress tensor for the Williamson fluid model reads as:

$$\overline{\mathbf{S}} = \left[\overline{\mu}_\infty - \left(\overline{\mu}_\infty - \overline{\mu}_0\right)\left(1 - \overline{\Gamma}\dot{\gamma}\right)^{-1}\right] \dot{\gamma}, \tag{6}$$

where $\overline{\mu}_\infty, \overline{\mu}_0$ the infinite and zero shear rate viscosity, $\overline{\Gamma}$ the time constant, and $\dot{\gamma}$ reads as:

$$\dot{\gamma} = \left(\frac{1}{2} \sum_m \sum_n \dot{\gamma}_{mn} \dot{\gamma}_{nm}\right)^{\frac{1}{2}} = \left(\frac{1}{2} S_t\right)^{\frac{1}{2}}, \tag{7}$$

where S_t is the second invariant strain tensor. For the present flow problem, we considered $\overline{\mu}_\infty = 0$ (the infinite shear rate viscosity is very small as compared to zero shear rate viscosity) and $\overline{\Gamma}\dot{\gamma} < i$ i.e. $i = 1$. Then, Equation (6) takes the following form:

$$\overline{\mathbf{S}} = \overline{\mu}_0 \left(1 - \overline{\Gamma}\dot{\gamma}\right)^{-1} \dot{\gamma}. \tag{8}$$

The energy equation to represent the heat exchange in the channel is as stated below. The law of conservation of energy in the dimensional mathematical pattern is given by:

$$S_h\left[D_{\bar{t}}+\overline{U}D_{\overline{X}}+\overline{V}D_{\overline{Y}}\right]T=\frac{K}{\rho}\left[D_{\overline{XX}}T+D_{\overline{YY}}T\right]+\frac{\overline{S}_{\overline{XY}}}{\rho}D_{\overline{Y}}\overline{U}. \tag{9}$$

In the above equation, S_h is the specific heat coefficient, K the thermal conductivity, and ρ the density of the governing fluid.

Introducing wave frame coordinates transformations with propagation velocity c away from the fixed frame read as:

$$\{\overline{x}+c\overline{t},\overline{u}+c,\overline{y},\overline{v},\overline{P}(\overline{x})\}=\{\overline{X},\overline{U},\overline{Y},\overline{V},\overline{P}\left(\overline{X},t\right)\} \tag{10}$$

Defining the dimensionless quantities as:

$$\begin{gathered}x=\frac{\overline{x}}{\lambda},y=\frac{\overline{y}}{\overline{d_i}},u=\frac{\overline{u}}{\lambda},v=\frac{\overline{v}}{\lambda},S_{xx}=\frac{\lambda}{\overline{\mu}_0c}\overline{S}_{\overline{XX}},S_{xy}=\frac{\overline{d_i}}{\overline{\mu}_0c}\overline{S}_{\overline{XY}},S_{yy}=\frac{\overline{d_i}}{\overline{\mu}_0c}\overline{S}_{\overline{YY}},\\
\theta=\frac{T-T_0}{T_i-T_0},p=\frac{\overline{d_i}^2}{c\lambda\overline{\mu}_0}P,\dot{\gamma}=\frac{\dot{\overline{\gamma}}\overline{d_1}}{c},a=\frac{\overline{a_i}}{\overline{d_i}},b=\frac{\overline{b_i}}{\overline{d_i}},d=\frac{\overline{d_j}}{\overline{d_i}},h_{i,j}=\frac{H_{i,j}}{\overline{d_i}},\end{gathered} \tag{11}$$

where θ is the dimensionless temperature profile.

By invoking the above transformations in Equations (4)–(6), we arrive at (after ignoring the bars):

$$\text{Re}\left[\delta uD_xu+vD_yu\right]=-D_xp+D_xS_{xx}+D_yS_{xy}, \tag{12}$$

$$\text{Re}\delta\left[\delta uD_xv+vD_yv\right]=-D_yp+D_yS_{yy}+\delta^2D_xS_{yx}, \tag{13}$$

$$P_r\text{Re}\left[\delta uD_x\theta+\delta vD_y\theta\right]=\left[-\delta^2D_{xx}\theta+D_{yy}\theta\right]+B_rS_{xy}D_yu\theta, \tag{14}$$

and

$$S_{xy}=-\left(1+We\dot{\gamma}\right)\left(D_yu+\delta D_xv\right), \tag{15}$$

where,

$$\delta=\frac{\overline{d_1}}{\lambda},\text{Re}=\frac{\rho c\overline{d_1}}{\mu_0},We=\frac{\Gamma c}{\overline{d_i}},B_r=P_rE_c,P_r=\frac{vS_h\rho}{K},E_c=\frac{c^2}{S_h\left(T_i-T_0\right)}. \tag{16}$$

In the above equation, We the Weissenberg number, E_c is the Eckert number, P_r the Prandlt number, Re the Reynolds number, and B_r the Brinkman number. Under the assumptions of long wavelength and low Reynolds numbers ($\delta\approx1$, Re ≈0), Equations (12)–(14) take the form:

$$D_xp=D_y\left[\left(1+WeD_yu\right)D_yu\right], \tag{17}$$

$$D_yp=0, \tag{18}$$

$$D_{yy}\theta=-B_r\left[\left(D_yu\right)^2+We\left(D_yu\right)^3\right]. \tag{19}$$

This equation implies that $p\neq p(y)$ so $\partial p/\partial x$ can be written as dp/dx. At $We=0$, the above equation turns into viscous fluid flow. The associated no slip and convective boundary conditions selected for the problem read as:

$$\begin{gathered}u=-1,\theta'-B_i\theta=-B_i\ \text{at}\ y=h_i\left(x\right)=1+a\cos2x\pi,\\
u=-1,\theta=0\ \text{at}\ y=h_j\left(x\right)=-d-b\cos\left(\phi+2\pi x\right).\end{gathered} \tag{20}$$

where B_i is the Biot number.

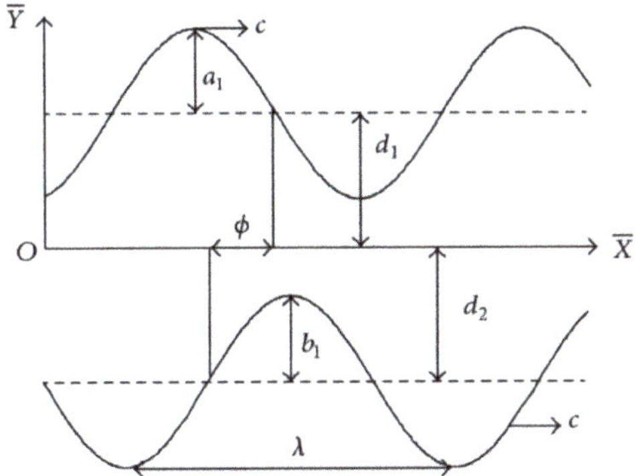

Figure 1. Flow structure.

3. Entropy Generation Analysis

According to the theory of thermodynamics, the physical process can be divided in to two types: Irreversible and reversible process. The characterization of such kind of procedures is associated with the change of entropy. Particularly, we say that the process is reversible if there is no change in the entropy, whereas, if the change occurs i.e., entropy is not zero, it shows that the process is irreversible. Therefore, the production of entropy is the measure of the irreversibility of a process. All the processes that arise in nature are irreversible and this reveals a significant obstacle in the study of that process.

The entropy generation in the dimensional form can be defined as:

$$S'_{gen} = \frac{K}{T_0^2} \left(D_{\overline{Y}} T \right)^2 + \frac{\overline{S}_{\overline{XY}}}{T_0} D_{\overline{Y}} \overline{U}. \tag{21}$$

Here we define some new dimensionless quantities in addition to those used above:

$$S'_g = \frac{K \left(T_i - T_0 \right)}{T_0^2 \overline{d_i^2}}, \Delta = \frac{T_0}{K \left(T_i - T_0 \right)}. \tag{22}$$

Using Equation (22) in Equation (21), we get the dimensionless form of entropy generation:

$$N = \frac{S'_{gen}}{S'_g} = \left(D_y \theta \right)^2 + \Delta B_r \left[-1 + We D_y u \right] \left(D_y u \right)^2. \tag{23}$$

In the above expression, Δ shows the entropy production characteristics and temperature difference parameter. Equation (23) is divided into two parts. The first is due to the finite temperature difference whereas the second part defines the fluid frictional irreversibility.

The Bejan number is describe as the entropy production ratio because of heat transfer irreversibility to the total entropy production:

$$Be = \frac{\left(D_y \theta \right)^2}{\left(D_y \theta \right)^2 + \Delta B_r \left[-1 + We D_y u \right] \left(D_y u \right)^2}. \tag{24}$$

Bejan number lies between 0 to 1. $Be < 1$ represents that the total entropy production dominates the total entropy production due to heat transfer. $Be = 1$ represents when the total entropy production is equal to entropy production due to heat transfer irreversibility.

4. Series Solution

Since Equation (17) is non linear, its exact solution may not be possible, therefore, we employ the regular perturbation method to find the solution. For perturbation solution, we expand u, F and dp/dx as:

$$u = \sum_{n=0}^{\infty} We^n u_n, \tag{25}$$

$$F = \sum_{n=0}^{\infty} We^n F_n, \tag{26}$$

$$\frac{dp}{dx} = \sum_{n=0}^{\infty} We^n \frac{dp_n}{dx}, \tag{27}$$

Substituting above expression in Equation (17) and their boundary conditions in Equation (20) and comparing the coefficients of powers of We we get the zeroth and first order systems which can be manipulated easily by a mathematical computing tool Mathematica and are conclusively stated as:

$$\begin{aligned} u = \frac{1}{2!}\left[-2 + C_1 h_1 h_2 - C_1 C_3 y + C_1 y^2\right] + \frac{1}{3!} We \Big[C_2 h_1 h_2 + C_1^2 C_3 h_1 h_2 \\ -C_2 (C_3 + y) y + C_1^2 \left(h_1^2 - 4h_1 h_2 - h_2^2\right) y + C_1^2 (3C_3 - 2y) y\Big] + O(We^2), \end{aligned} \tag{28}$$

$$\frac{dp}{dx} = 12C_1 + 12WeC_2 + O(We^2), \tag{29}$$

where the constant are defined as:

$$C_1 = \frac{12\,(1 + d - h_1 + h_2 - Q)}{(h_1 - h_2)^3}, \tag{30}$$

$$C_2 = \frac{36\,(1 + d - Q)}{(h_1 - h_2)^3}, \tag{31}$$

$$C_3 = h_1 + h_2, \tag{32}$$

$$C_4 = C_3^2 + h_1 h_2, \tag{33}$$

$$C_5 = 17C_3^2 + 4h_1 h_2, \tag{34}$$

$$C_6 = 7C_3^2 + 2h_1 h_2. \tag{35}$$

The solution for velocity u obtained by above perturbation method can be used in Equation (19). The final solution for θ can be obtained by integrating Equation (19) along with their associated boundary conditions (See Equation (20)) and can be written as:

$$\theta = \theta_1 + \theta_2 y + \theta_3 y^2 + \theta_4 y^3 + \theta_5 y^4 + \theta_6 y^5 + \theta_7 y^6 + \theta_8 y^7 + \theta_9 y^8, \tag{36}$$

where constants of integration θ_1 and θ_2 can be evaluated by using boundary conditions defined in Equation (20) and the expression obtained are very large and therefore are not presented here. The remaining constants are defined as:

$$\theta_3 = \frac{B_r\sqrt{C_3}}{432}\left[3\frac{dp_0}{dx} + \left(C_3\left(\frac{dp_0}{dx}\right)^2 + 3\frac{dp_1}{dx}\right)We\right]^2 \times \left[-6 + 3C_3We\frac{dp_0}{dx} + C_3\left(C_3\left(\frac{dp_0}{dx}\right)^2 + 3\frac{dp_1}{dx}\right)We^2\right], \tag{37}$$

$$\theta_4 = -\frac{B_rC_3}{72}\left(\frac{dp_0}{dx} + C_3We\left(\frac{dp_0}{dx}\right)^2 + \frac{dp_1}{dx}\right) \times \left[3\frac{dp_0}{dx} + \left(C_3\left(\frac{dp_0}{dx}\right)^2 + 3\frac{dp_1}{dx}\right)We\right] \times \left[-4 + 3C_3We\frac{dp_0}{dx} + C_3\left(C_3\left(\frac{dp_0}{dx}\right)^2 + 3\frac{dp_1}{dx}\right)We^2\right], \tag{38}$$

$$\theta_5 = \frac{B_r}{12}\left[-12\left(\frac{dp_0}{dx}\right)^2 - 6\frac{dp_0}{dx}\left(3C_3\left(\frac{dp_0}{dx}\right)^2 + 4\frac{dp_1}{dx}\right) + \left((7h_1+5h_2)(5h_1+7h_2)\left(\frac{dp_0}{dx}\right)^4 + 18C_3\left(\frac{dp_0}{dx}\right)^2\frac{dp_1}{dx} - 12\left(\frac{dp_1}{dx}\right)^2\right)We^2 + 6\frac{dp_0}{dx}\left(6C_3C_4\left(\frac{dp_0}{dx}\right)^4 + C_5\left(\frac{dp_0}{dx}\right)^2\frac{dp_1}{dx} + 9C_3\left(\frac{dp_1}{dx}\right)^2\right) + C_3\left(C_3\left(\frac{dp_0}{dx}\right)^2 + 3\frac{dp_1}{dx}\right)\left(C_6\left(\frac{dp_0}{dx}\right)^4 + 15C_3\left(\frac{dp_0}{dx}\right)^2\frac{dp_1}{dx} + 6\left(\frac{dp_1}{dx}\right)^2\right)We^4\right], \tag{39}$$

$$\theta_6 = -\frac{B_rWe}{20}\left(\frac{dp_0}{dx} + C_3\left(\frac{dp_0}{dx}\right)^2 + We\frac{dp_1}{dx}\right) \times \left[-\left(\frac{dp_0}{dx}\right)^2 + \frac{dp_0}{dx}We\left(5C_3\left(\frac{dp_0}{dx}\right)^2 + 2\frac{dp_1}{dx}\right)We + \left(2C_4\left(\frac{dp_0}{dx}\right)^4 + 5C_3\left(\frac{dp_0}{dx}\right)^2\frac{dp_1}{dx} + \left(\frac{dp_1}{dx}\right)^2\right)We^2\right], \tag{40}$$

$$\theta_7 = \frac{B_rWe^2}{60}\left(\frac{dp_0}{dx}\right)^2\left[4\left(\frac{dp_0}{dx}\right)^4 + 3\frac{dp_0}{dx}\left(5C_3\left(\frac{dp_0}{dx}\right)^2 + 4\frac{dp_1}{dx}\right)We + \left(C_6\left(\frac{dp_0}{dx}\right)^4 + 15C_3\left(\frac{dp_0}{dx}\right)^2\frac{dp_1}{dx} + 6\left(\frac{dp_1}{dx}\right)^2\right)We^2\right], \tag{41}$$

$$\theta_8 = -\frac{B_rWe^3}{14}\left(\frac{dp_0}{dx}\right)^4\left[\frac{dp_0}{dx} + C_3We\left(\frac{dp_0}{dx}\right)^2 + We\frac{dp_1}{dx}\right], \tag{42}$$

$$\theta_9 = \frac{1}{56} B_r We^4 \left(\frac{dp_0}{dx}\right)^6. \tag{43}$$

The dimensionless mean flow reads as:

$$F = d + 1 - Q. \tag{44}$$

and

$$F = \int_{h_2}^{h_1} u dy. \tag{45}$$

The expression for entropy generation and Bejan number can be easily obtained by incorporating value of u and θ in Equation (24).

5. Discussion

In this section, we present our results by varying the quantities under the variation of several factors. Figures of temperature profile θ, entropy generation coefficient N, and streamlines are illustrated below. Figures 2–5 reflect the behavior of θ for some useful parameters. Entropy generation graphs are given in Figures 6–11. The streamlines conducting the flow samples are depicted in Figures 12 and 13.

Figure 2 shows the impact of parameters a and b on temperature profile θ. It can be observed from this plot that temperature is getting increased for both parameters from the lower wall to the upper wall. Figure 3 shows the mechanism of the Biot number and Brinkman number. Biot number is an important mechanism to determine the heat transfer. It can be visualized from this figure that an enhancement in Biot number tends to boost the temperature profile while the contrary behavior has been observed with the Brinkman number. Brinkman number is the product of Eckert and Prandtl numbers $B_r = P_r E_c$, or it is the ratio of the heat generated by viscous dissipation and propagation of heat by molecular conduction, such as, the ratio of the viscous heat production to extrinsic heating. Therefore, the enhancement of Brinkman's number tends to increase the temperature profile. It can be seen in Figure 4 that the volumetric flow rate significantly enhances the temperature profile. It can also be noticed that the temperature profile has a lower magnitude for smaller values of d whereas the behavior is converse for higher values. It can be viewed from Figure 5 that the Weissenberg number causes a remarkable resistance for higher values. By enhancing the Weissenberg number, the elastic forces are more dominant, which diminishes the temperature profile. However, the phase difference ϕ also produces a significant resistance in the temperature profile.

Figures 6–9 are presented for entropy profiles against the leading parameters. It can be viewed from Figure 6 that an increment in a and b tends to boost the entropy profile whereas the entropy profile is increasing along the whole channel. Figure 7 shows that by increasing the Brinkman number, the entropy profile rises, and it decreases by increasing the Weissenberg number. However, the entropy remains positive and growing along the entire channel. It is seen from Figure 8 that the Biot number enhances the entropy profile. It can be seen that at the lower wall, the entropy profile is maximum and minimum at the upper wall, whereas it is uniform in the middle of the channel. The entropy profile for various values of Δ is presented in Figure 9. It is noticed in this figure that the entropy profile is uniform, and no change occurs in the middle of the channel i.e., $y \in (0, 0.5)$. Although it shows a decreasing pattern, but it rises along the upper wall of the channel and remains positive.

Figures 10 and 11 are plotted for the Bejan number profile against the governing parameters. It is observed from Figure 10 that the Bejan number profile diminishes for higher values of the Brinkman number and shows a converse behavior for the Weissenberg number. In Figure 11, we can see that the phase difference shows versatile behavior for higher values on the Bejan number profile. When Bejan number rises, then the phase difference's effects are negligible for the domain $y \in (0, 1.3)$, while when the Bejan number is small, it decreases in a similar area.

The most interesting and useful phenomena of peristaltic motion are trapping, which is plotted in Figures 12 and 13 via streamlines. It was found that by enhancing the phase difference parameter, the effects are negligible on the trapping bolus despite the fact that an unusual movement in the magnitude of the bolus is noticed. Furthermore, we can see in Figure 13 that an increment in the Weissenberg number profile tends to diminish the width of the trapping bolus. The number of boluses disappeared more quickly in the lower region as compared with the upper one.

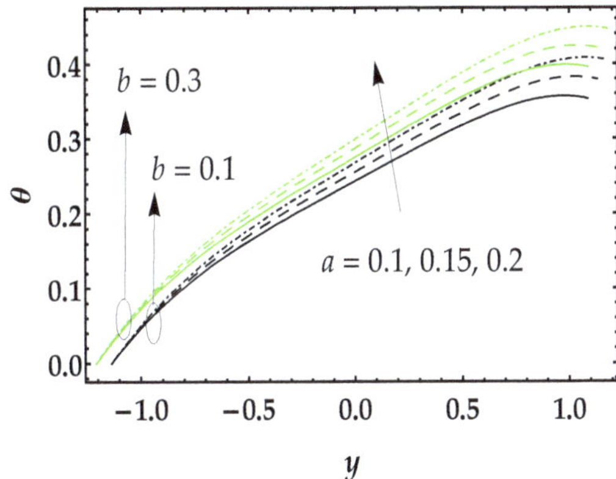

Figure 2. Temperature distribution for different values of a and b. Solid line: $a = 0.1$, dashed line: $a = 0.15$ and dot-dashed line: $a = 0.2$.

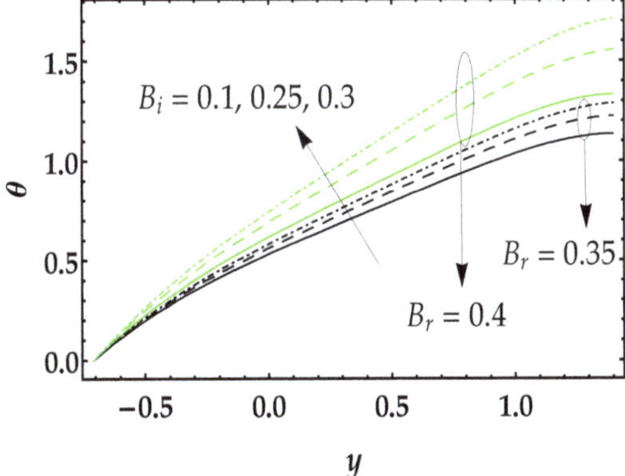

Figure 3. Temperature distribution for different values of B_i and B_r. Solid line: $B_i = 0.1$, dashed line: $B_i = 0.25$ and dot-dashed line: $B_i = 0.3$.

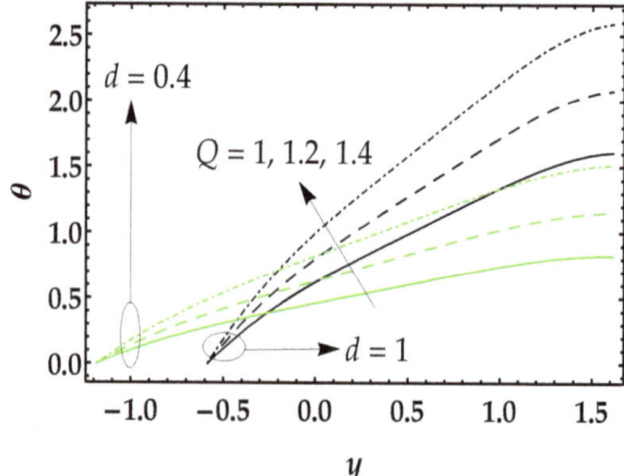

Figure 4. Temperature distribution for different values of Q and d. Solid line: $Q = 1.0$, dashed line: $Q = 1.2$ and dot-dashed line: $Q = 1.4$.

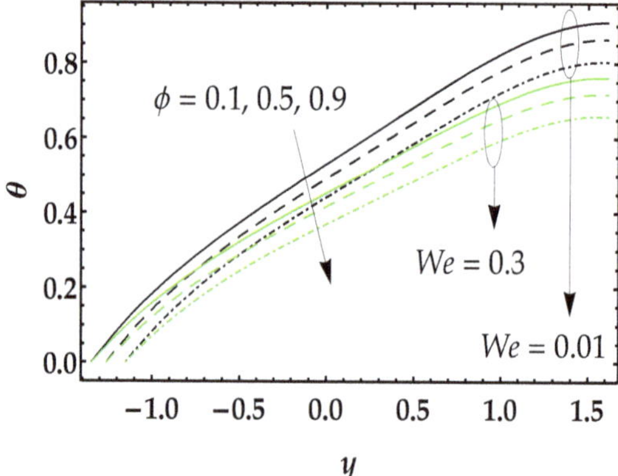

Figure 5. Temperature distribution for different values of ϕ and We. Solid line: $\phi = 0.1$, dashed line: $\phi = 0.5$ and dot-dashed line: $\phi = 0.9$.

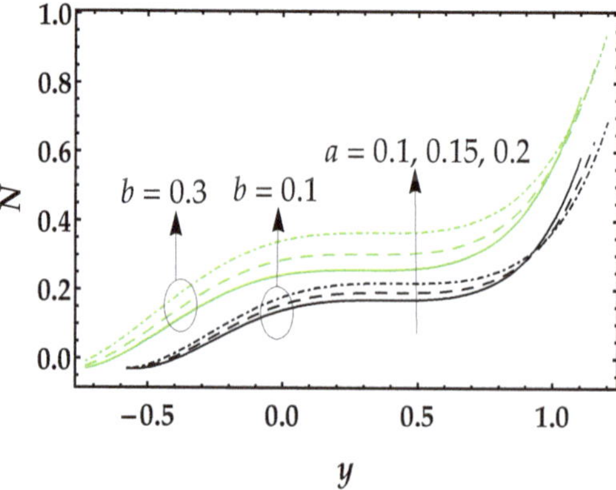

Figure 6. Entropy profile for different values of a and b. Solid line: $a = 0.1$, dashed line: $a = 0.15$, and dot-dashed line: $a = 0.2$.

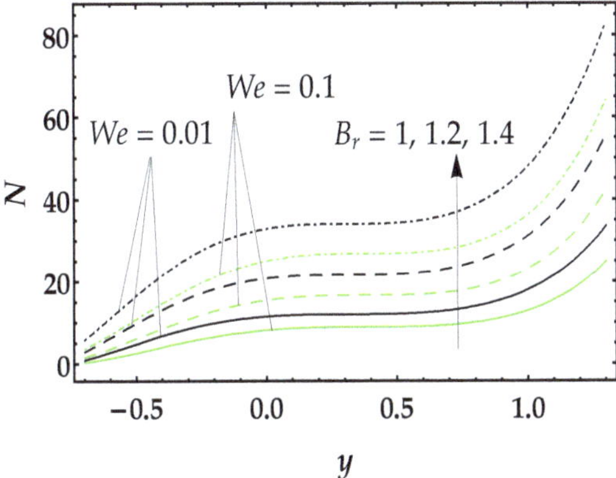

Figure 7. Entropy profile for different values of We and B_r. Solid line: $B_r = 1.0$, dashed line: $B_r = 1.2$, and dot-dashed line: $B_r = 1.4$.

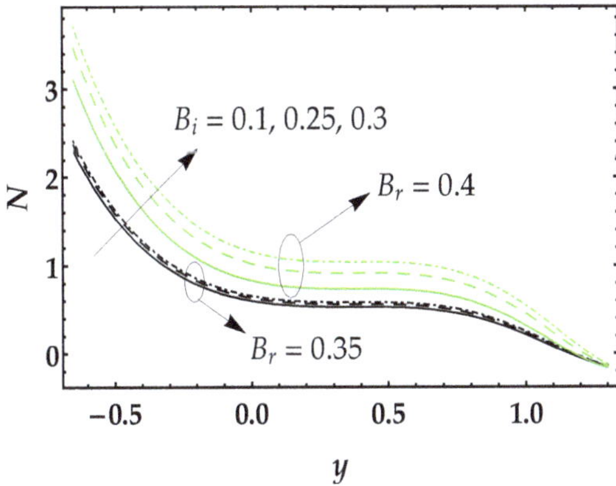

Figure 8. Entropy profile for different values of B_i and B_r. Solid line: $B_i = 0.1$, dashed line: $B_i = 0.25$, and dot-dashed line: $B_i = 0.3$.

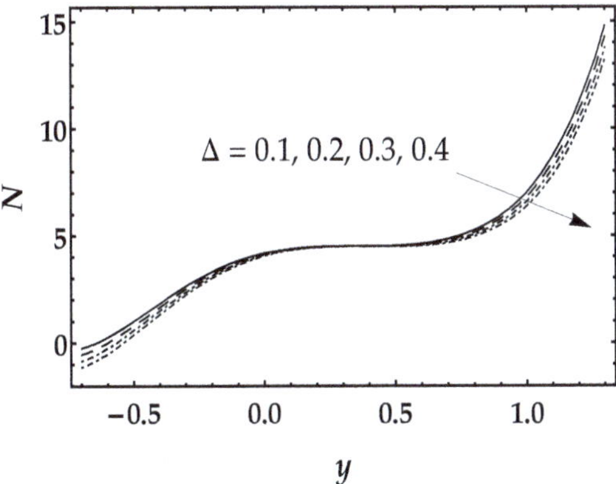

Figure 9. Entropy profile for different values of Δ. Solid line: $\Delta = 0.1$, dashed line: $\Delta = 0.2$, dot-dashed line: $\Delta = 0.3$, and dot-dashed line: $\Delta = 0.4$.

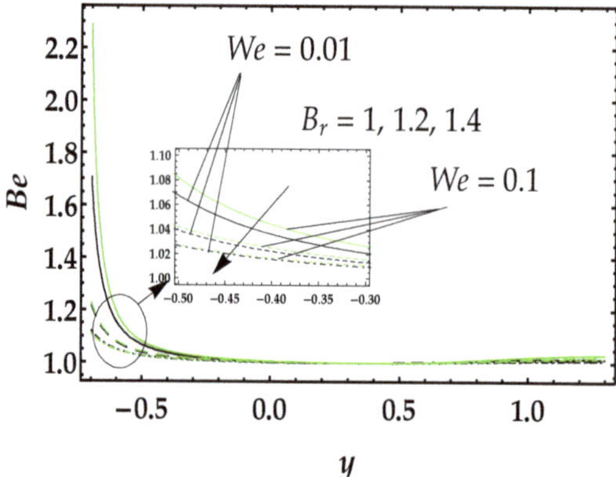

Figure 10. Bejan number for different values of We and B_r. Solid line: $B_r = 1.0$, dashed line: $B_r = 1.2$, and dot-dashed line: $B_r = 1.4$.

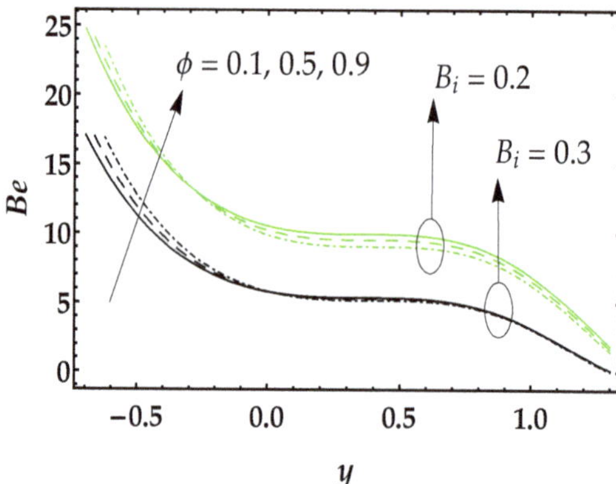

Figure 11. Bejan number for different values of ϕ and B_i. Solid line: $\phi = 0.1$, dashed line: $\phi = 0.5$, and dot-dashed line: $\phi = 0.9$.

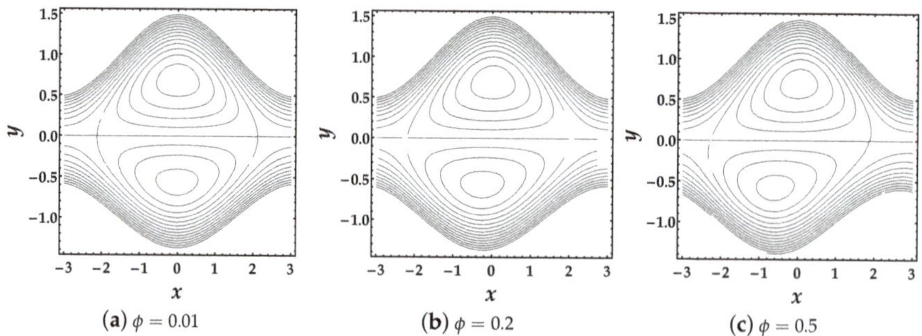

Figure 12. Trapping mechanism for different values of ϕ.

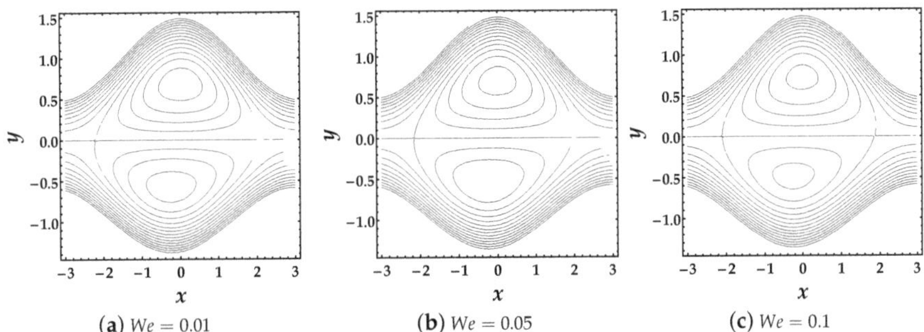

Figure 13. Trapping mechanism for different values of We.

6. Conclusions

In this study, we analyzed the entropy generation on the asymmetric peristaltic propulsion of non-Newtonian fluid with convective boundary conditions. The Williamson fluid model was considered to examine the entropy profile. The mathematical modeling was performed under the approximation of small Reynolds number and long wavelength of the peristaltic wave. A regular perturbation method was employed to get the series solutions up to third-order approximation. The significant results of the governing flow problem are summarized below:

(i) It was noticed that the temperature profile revealed an increasing behavior by increasing the amplitude in the upper and lower region;
(ii) The Biot number and Brinkman number significantly enhanced the temperature profile, whereas the behavior is converse for the phase difference parameter and Weissenberg number;
(iii) Entropy profile represented an increment profile for higher values of Brinkmann number and Biot number, and a decrement behavior for the Weissenberg number;
(iv) The Weissenberg number boosedt the Bejan number profile, whereas it decreased due to the Biot number and Brinkman number;
(v) Trapping mechanism showed that the phase difference parameter affected the magnitude of the trapped bolus, while the Weissenberg number not only affected the magnitude of the trapped bolus and the number of trapped boluses reduced in the lower region;
(vi) The non-Newtonian results in the present study could be reduced to Newtonian fluid flow by taking $We = 0$.

The present results provide an excellent benchmark for further study on the entropy production with mass transfer and peristaltic pumping mechanism. The mass transfer phenomena with magnetic and porosity effects that were not covered in this paper is a topic for future research.

Author Contributions: Investigation, A.R.; M.M.B., Methodology; Conceptualization, R.E.; Validation, A.Z.; Writing—review & editing, S.M.S. All authors have read and agreed to the published version of the manuscript.

Funding: This research received no external funding.

Conflicts of Interest: The authors declare no conflict of interest.

Nomenclature

$\overline{d_1}, \overline{d_2}$	channel width
T	temperature
c	wave speed
$\overline{t}$	time
$\overline{X}, \overline{Y}$	coordinate system
$\overline{U}, \overline{V}$	velocity components
$\overline{\mathbf{S}}$	stress tensor
S_h	specific heat
K	thermal conductivity
P	pressure
a, b	wave amplitude
Re	Reynold's number
E_c	Eckert number
P_r	Prandtl number
B_r	Brinkmann number
B_i	Biot number
We	Weissenberg number
Be	Bejan number
S'_{gen}	entropy

Greek Symbol

ϕ	phase difference
ρ	density
λ	wavelength
μ	viscosity
$\overline{\Gamma}$	time constant
δ	wave number

References

1. Goldstein, M.; Goldstein, I.F. *The Refrigerator and the Universe: Understanding the Laws of Energy*; Harvard University Press: Cambridge, MA, USA, 1995.
2. Kleiber, M. *The Fire of Life: An Introduction to Animal Energetics*; Robert E Kreiger Co.: New York, NY, USA, 1975.
3. Holmes, F.L. *Lavoisier and the Chemistry of Life: An Exploration of Scientific Creativity*; Number 4; Univ. of Wisconsin Press: Madison, WI, USA, 1985.
4. Silva, C.; Annamalai, K. Entropy generation and human aging: Lifespan entropy and effect of physical activity level. *Entropy* **2008**, *10*, 100–123. [CrossRef]
5. Van Dinh, Q.; Vo, T.Q.; Kim, B. Viscous heating and temperature profiles of liquid water flows in copper nanochannel. *J. Mech. Sci. Technol.* **2019**, *33*, 3257–3263. [CrossRef]
6. Ghorbanian, J.; Celebi, A.T.; Beskok, A. A phenomenological continuum model for force-driven nano-channel liquid flows. *J. Chem. Phys.* **2016**, *145*, 184109. [CrossRef]
7. Silva, C.A.; Annamalai, K. Entropy generation and human aging: Lifespan entropy and effect of diet composition and caloric restriction diets. *J. Thermodyn.* **2009**, *2009*, 186723. [CrossRef]
8. Hershey, D. *Lifespan and Factors Affecting It: Aging Theories in Gerontology*; Charles C. Thomas Publisher: Springfield, MA, USA, 1974.
9. Hershey, D.; Wang, H.H. *A New Age-Scale for Humans*; Lexington Books: Lexington, KY, USA; Toronto, ON, Canada, 1980.
10. Rahman, M.A. A novel method for estimating the entropy generation rate in a human body. *Therm. Sci.* **2007**, *11*, 75–92. [CrossRef]
11. Aoki, I. Entropy flow and entropy production in the human body in basal conditions. *J. Theor. Biol.* **1989**, *141*, 11–21. [CrossRef]

12. Annamalai, K.; Puri, I.K.; Jog, M.A. *Advanced Thermodynamics Engineering*; CRC Press: Boca Raton, FL, USA, 2011.
13. Bejan, A. Constructal theory: From thermodynamic and geometric optimization to predicting shape in nature. *Energy Convers. Manag.* **1998**, *39*, 1705–1718. [CrossRef]
14. Bejan, A. The tree of convective heat streams: Its thermal insulation function and the predicted 3/4-power relation between body heat loss and body size. *Int. J. Heat Mass Transf.* **2001**, *44*, 699–704. [CrossRef]
15. Rashidi, M.; Kavyani, N.; Abelman, S. Investigation of entropy generation in MHD and slip flow over a rotating porous disk with variable properties. *Int. J. Heat Mass Transf.* **2014**, *70*, 892–917. [CrossRef]
16. Komurgoz, G.; Arikoglu, A.; Ozkol, I. Analysis of the magnetic effect on entropy generation in an inclined channel partially filled with a porous medium. *Numer. Heat Transf. Part A Appl.* **2012**, *61*, 786–799. [CrossRef]
17. Rashidi, M.; Keimanesh, M.; Bég, O.A.; Hung, T. Magnetohydrodynamic biorheological transport phenomena in a porous medium: A simulation of magnetic blood flow control and filtration. *Int. J. Numer. Methods Biomed. Eng.* **2011**, *27*, 805–821. [CrossRef]
18. Akbar, N.S.; Raza, M.; Ellahi, R. Peristaltic flow with thermal conductivity of H2O+ Cu nanofluid and entropy generation. *Results Phys.* **2015**, *5*, 115–124. [CrossRef]
19. Rashidi, M.; Bhatti, M.; Abbas, M.; Ali, M. Entropy generation on MHD blood flow of nanofluid due to peristaltic waves. *Entropy* **2016**, *18*, 117. [CrossRef]
20. Akbar, N.; Raza, M.; Ellahi, R. Endoscopic Effects with Entropy Generation Analysis in Peristalsis for the Thermal Conductivity of 2 HO Nanofluid. *J. Appl. Fluid Mech.* **2016**, *9*, 1721–1730.
21. Abbas, M.; Bai, Y.; Rashidi, M.; Bhatti, M. Analysis of entropy generation in the flow of peristaltic nanofluids in channels with compliant walls. *Entropy* **2016**, *18*, 90. [CrossRef]
22. Bhatti, M.M.; Abbas, M.A.; Rashidi, M.M. Entropy Generation for Peristaltic Blood Flow with Casson Model and Consideration of Magnetohydrodynamics Effects. *Walailak J. Sci. Technol. (WJST)* **2017**, *14*, 451–461.
23. Ranjit, N.; Shit, G. Entropy generation on electro-osmotic flow pumping by a uniform peristaltic wave under magnetic environment. *Energy* **2017**, *128*, 649–660. [CrossRef]
24. Majee, S.; Shit, G. Numerical investigation of MHD flow of blood and heat transfer in a stenosed arterial segment. *J. Magn. Magn. Mater.* **2017**, *424*, 137–147. [CrossRef]
25. Bhatti, M.; Sheikholeslami, M.; Zeeshan, A. Entropy analysis on electro-kinetically modulated peristaltic propulsion of magnetized nanofluid flow through a microchannel. *Entropy* **2017**, *19*, 481. [CrossRef]
26. Ellahi, R.; Zeeshan, A.; Hussain, F.; Asadollahi, A. Peristaltic blood flow of couple stress fluid suspended with nanoparticles under the influence of chemical reaction and activation energy. *Symmetry* **2019**, *11*, 276. [CrossRef]
27. Prakash, J.; Tripathi, D.; Tiwari, A.K.; Sait, S.M.; Ellahi, R. Peristaltic Pumping of Nanofluids through a Tapered Channel in a Porous Environment: Applications in Blood Flow. *Symmetry* **2019**, *11*, 868. [CrossRef]
28. Mekheimer, K.S.; Zaher, A.; Abdellateef, A. Entropy hemodynamics particle-fluid suspension model through eccentric catheterization for time-variant stenotic arterial wall: Catheter injection. *Int. J. Geom. Methods Mod. Phys.* **2019**, *16*, 1950164. [CrossRef]

Article

Serious Solutions for Unsteady Axisymmetric Flow over a Rotating Stretchable Disk with Deceleration

Muhammad Adil Sadiq

Department of Mathematics, DCC-KFUPM, KFUPM Box 5084, Dhahran 31261, Saudi Arabia; adilsadiq@kfupm.edu.sa; Tel.: +966-13-868-3300

Received: 12 December 2019; Accepted: 2 January 2020; Published: 3 January 2020

Abstract: In this article, the author has examined the unsteady flow over a rotating stretchable disk with deceleration. The highly nonlinear partial differential equations of viscous fluid are simplified by existing similarity transformation. Reduced nonlinear ordinary differential equations are solved by homotopy analysis method (HAM). The convergence of HAM solutions is also obtained. A comparison table between analytical solutions and numerical solutions is also presented. Finally, the results for useful parameters, i.e., disk stretching parameters and unsteadiness parameters, are found.

Keywords: homotopy analysis method; rotating stretchable disk; newtonian fluid; axisymmetric flow

1. Introduction

Recently, due to the massive practical application in the scientific and technical field, the study of the rotating stretchable disk has become significant, such as thermal power generation system, medical equipment's, computer storage devices, rotating machinery, gas turbine routers, air cleaning machines, crystal growth process, and in aerodynamic applications [1]. Initially, von Kármán [2] conducted a study on rotating disk. Several researchers then illustrated the different aspects of this important analysis. Fang and Zhang [3] have highlighted the flow between two stretchable disks and found the exact solutions. The parameters analysis and optimization of entropy generation in unsteady magneto hydrodynamics flow over a rotating stretchable flow over a rotating disk using artificial neural network and practical swarm optimization algorithm was presented by Rashidi et al. [4]. Recently, Fang and Tao [5] wrote about the unsteady flow over a rotating stretchable disk with deceleration. After using the similarity analysis, they found the numerical solutions.

In many situations, exact solutions are very difficult and in most of the cases exact solutions are impossible. Therefore, series solutions are more useful if they satisfy the given initial and boundary value problems. Nevertheless, there are various analytical approaches, and each approach has certain limitations. However, homotopy analysis method (HAM) has many advantages over many analytical methods. Liao [6] introduced the idea of HAM, which is used by many researchers effectively. Some useful studies are cited in [7–15]. The purpose of this article is to illustrate the application of HAM for unsteady Newtonian fluid flow over a rotating stretchable disk with declaration. Tables provide a correlation between current HAM solution and Fang and Tao's [5] numerical solution.

2. Formulation of the Problem

Let us consider an incompressible, laminar, and unsteady flow of a viscous fluid or Newtonian fluid over a stretchable disk, which is rotating about the z-axis with time dependent angular velocity $\Omega\prime(t) = \frac{\Omega}{1-bt}$, where Ω is constant angular speed of the disk and 'b' is the measure of unsteadiness. Flow is due to the rotation of the stretchable disk and is axisymmetric about the z-axis. Figure 1 describe the geometry of the proposed problem. The governing equations for an unsteady three-dimensional flow of viscous fluid in cylindrical coordinates are shown below.

$$\frac{1}{r}\frac{\partial}{\partial r}(ru) + \frac{\partial w}{\partial z} = 0, \tag{1}$$

$$\frac{\partial u}{\partial t} + u\frac{\partial u}{\partial r} + w\frac{\partial u}{\partial z} - \frac{v^2}{r} = -\frac{1}{\rho}\frac{\partial p}{\partial r} + v\left(\frac{1}{r}\frac{\partial}{\partial r}(r\tau_{rr}) + \frac{\partial}{\partial z}(\tau_{zr}) - \frac{\tau_{\theta\theta}}{r}\right), \tag{2}$$

$$\frac{\partial v}{\partial t} + u\frac{\partial v}{\partial r} + w\frac{\partial v}{\partial z} - \frac{uv}{r} = v\left(\frac{\partial}{\partial r}(\tau_{r\theta}) + \frac{\partial}{\partial z}(\tau_{z\theta}) + \frac{\tau_{\theta r} + \tau_{r\theta}}{r}\right), \tag{3}$$

$$\frac{\partial w}{\partial t} + u\frac{\partial w}{\partial r} + w\frac{\partial w}{\partial z} = -\frac{1}{\rho}\frac{\partial p}{\partial z} + v\left(\frac{1}{r}\frac{\partial}{\partial r}(r\tau_{rz}) + \frac{\partial}{\partial z}(\tau_{zz})\right), \tag{4}$$

where r is along the radial direction, θ is along the azimuthal direction, and z is in normal direction to the axis. Here, Equation (1) is the continuity equation and Equations (2) and (4) represent the momentum equation for incompressible flow.

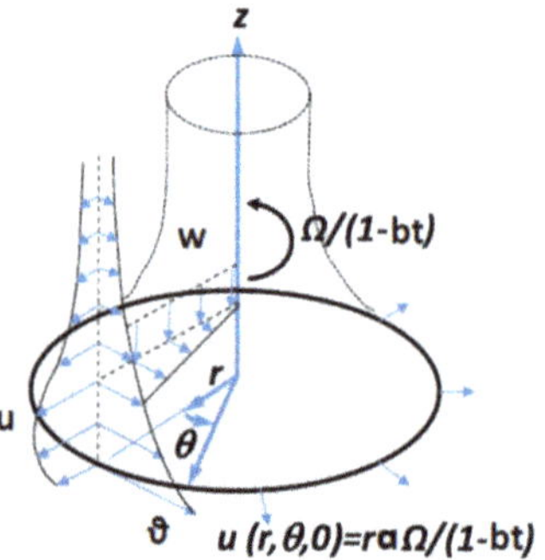

Figure 1. Geometry of the problem.

Where $u,\ v$, and w are the velocities along r, θ, and z directions, ρ is the density of fluid, p is the pressure, v is the kinematic viscosity, and $\tau_{rr},\ \tau_{r\theta},\ \tau_{zr},\ \tau_{z\theta},\ \tau_{zz}$ are the stress which are defined as

$$\begin{bmatrix} \tau_{rr} & \tau_{r\theta} & \tau_{rz} \\ \tau_{\theta r} & \tau_{\theta\theta} & \tau_{\theta z} \\ \tau_{zr} & \tau_{z\theta} & \tau_{zz} \end{bmatrix} = \begin{bmatrix} 2\frac{\partial u}{\partial r} & \frac{\partial v}{\partial r} - \frac{v}{r} & \frac{\partial u}{\partial z} + \frac{\partial w}{\partial r} \\ \frac{\partial v}{\partial r} - \frac{v}{r} & \frac{2u}{r} & \frac{\partial v}{\partial z} \\ \frac{\partial u}{\partial u} + \frac{\partial w}{\partial r} & \frac{\partial v}{\partial z} & 2\frac{\partial w}{\partial z} \end{bmatrix}. \tag{5}$$

The proposed boundary conditions are specified in accordance with the geometry of the problem as

$$\begin{gathered} u(r,\theta,0) = \tfrac{a\Omega r}{1-bt},\ v(r,\theta,0) = \tfrac{\Omega r}{1-bt},\ w(r,\theta,0) = 0, \\ u(r,\ \theta,\ \infty) = v(r,\theta,\infty) = 0, \end{gathered} \tag{6}$$

where 'a' is the disk stretching parameter.

Introducing, the similarity transformation used in [5] are

$$\begin{gathered} u = \tfrac{\Omega r}{1-bt} f'(\eta),\ v = \tfrac{\Omega r}{1-bt} g(\eta),\ w = -2\tfrac{\sqrt{\Omega v}}{\sqrt{1-bt}} f(\eta), \\ p = \tfrac{\rho v \Omega}{1-bt} P(\eta),\ \text{and}\ \eta = \sqrt{\tfrac{\Omega}{v}}\tfrac{z}{\sqrt{1-bt}}. \end{gathered} \tag{7}$$

Applying these similarities into the above equations, following non-dimensional equations along with boundary conditions can be obtained as

$$f''' + 2ff'' - f'^2 + g^2 = S\left(\frac{\eta}{2}f'' + f'\right), \tag{8}$$

$$g'' - 2f'g - 2fg' = S\left(\frac{\eta}{2}g\prime + g\right), \tag{9}$$

$$P' = 2f'' + 4ff'' - S(\eta f' + f), \tag{10}$$

$$f(0) = 0,\ f'(0) = a,\ g(0) = 1,\ f' \to 0 \ and\ g \to 0,\ as\ \eta \to \infty, \tag{11}$$

where $S = b/\Omega$ is the unsteadiness parameter.

3. Homotopy Analysis Method

Homotopy Analysis Method (HAM) [6–15] is used to find an analytical solution to Equations (8) and (11). The velocity distribution $f(\eta)$ and $g(\eta)$ can be expressed by a set of base functions

$$\left\{\eta^n exp^{(-m\eta)}\middle|m, n \geq 0\right\} \tag{12}$$

in the form

$$f(\eta) = a_{0,0}^0 + \sum_{n=0}^{\infty}\sum_{k=0}^{\infty} a_{m,n}^k \eta^n \exp^{(-m\eta)}, \tag{13}$$

$$g(\eta) = b_{0,0}^0 + \sum_{n=0}^{\infty}\sum_{k=0}^{\infty} b_{m,n}^k \eta^n \exp^{(-m\eta)} \tag{14}$$

in which $a_{m,n}^k$ and $b_{m,n}^k$ are the coefficients, the initial guesses f_0 and g_0 can be selected on the basis of the law of the solution expressions and of the boundary conditions:

$$f_0(\eta) = a(1 - \exp^{-\eta}), \tag{15}$$

$$g_0(\eta) = \exp^{-\eta}. \tag{16}$$

The auxiliary linear operators are

$$\mathcal{L}_1 = \frac{d^3}{d\eta^3} + \frac{d^2}{d\eta^2}, \tag{17}$$

$$\mathcal{L}_2 = \frac{d^2}{d\eta^2} + \frac{d}{d\eta}, \tag{18}$$

which satisfy

$$\mathcal{L}_1[C_1 + C_2 \exp^{-\eta} + C_3\eta] = 0, \tag{19}$$

$$\mathcal{L}_2[C_4 \exp^{-\eta} + C_5] = 0, \tag{20}$$

where $C_i(i = 1 - 5)$ are integral constants.

3.1. Zeroth-Order Deformation Equation

If $q \in [0, 1]$ denote an embedding parameter, $\hbar_f$ and $\hbar_g$ indicate the non zero auxiliary parameters for $f(\eta)$ and $g(\eta)$, the zeroth-order deformations for the given problem are

$$(1 - q)_f\left[\hat{f}(\eta; q) - f_0(\eta)\right] = q\hbar_f N_f\left[\hat{f}(\eta; q), \hat{g}(\eta; q)\right], \tag{21}$$

$$(1 - q)_g[\hat{g}(\eta; q) - g_0(\eta)] = q\hbar_g N_g\left[\hat{g}(\eta; q), \hat{f}(\eta; q)\right], \tag{22}$$

$$\hat{f}(0;q) = 0,\ \hat{f}'\ (0;q) = a,\ \hat{g}(0;q) = 1\ \hat{f}'\ (\infty;q) = \hat{g}(\infty;q) = 0. \tag{23}$$

Defining the nonlinear operators for the above problem as

$$\begin{aligned} N_f\left[\hat{f}(\eta;q), \hat{g}(\eta;q)\right] = & \frac{\partial^3 \hat{f}(\eta;q)}{\partial \eta^3} + 2\hat{f}(\eta;q)\frac{\partial^2 \hat{f}(\eta;q)}{\partial \eta^2} - \left(\frac{\partial \hat{f}(\eta;q)}{\partial \eta}\right)^2 \\ & +(\hat{g}(\eta;q))^2 - S\left(\frac{\eta}{2}\frac{\partial^2 \hat{f}(\eta;q)}{\partial \eta^2} + \frac{\partial \hat{f}(\eta;q)}{\partial \eta}\right), \end{aligned} \tag{24}$$

$$\begin{aligned} N_g\left[\hat{g}(\eta;q), \hat{f}(\eta;q)\right] = & \frac{\partial^2 \hat{g}(\eta;q)}{\partial \eta^2} + 2\hat{f}(\eta;q)\frac{\partial \hat{g}(\eta;q)}{\partial \eta} \\ & -2\hat{g}(\eta;q)\frac{\partial \hat{f}(\eta;q)}{\partial \eta} - s\left(\frac{\eta}{2}\frac{\partial \hat{g}(\eta;q)}{\partial \eta} + \hat{g}(\eta;q)\right). \end{aligned} \tag{25}$$

For $q = 0$ and $q = 1$, one can have

$$\hat{f}(\eta;0) = f_0(\eta),\ \hat{f}(\eta:1) = f(\eta), \tag{26}$$

$$\hat{g}(\eta;0) = g_0(\eta),\ \hat{g}(\eta:1) = g(\eta). \tag{27}$$

By Taylor's theorem

$$\hat{f}(\eta;q) = f_0(\eta) + \sum_{m=1}^{\infty} f_m(\eta)q^m,\ f_m(\eta) = \frac{1}{m!}\frac{\partial^m f(\eta;q)}{\partial \eta^m}\Bigg|_{q=0}, \tag{28}$$

$$\hat{g}(\eta;q) = g_0(\eta) + \sum_{m=1}^{\infty} g_m(\eta)q^m,\ g_m(\eta) = \frac{1}{m!}\frac{\partial^m g(\eta;q)}{\partial \eta^m}\Bigg|_{q=0}, \tag{29}$$

and

$$f(\eta) = f_0(\eta) + \sum_{m=1}^{\infty} f_m(\eta), \tag{30}$$

$$g(\eta) = g_0(\eta) + \sum_{m=1}^{\infty} g_m(\eta). \tag{31}$$

3.2. Mth-Order Deformation

Differentiating the zeorth-order deformation Equations (21) and (23) with respect to q, then setting $q = 0$, and finally dividing them by $m!$, the mth-order deformation equations can be obtained as

$$\mathcal{L}_1[f_m(\eta) - \chi_m f_{m-1}(\eta)] = \hbar_f R_m^f(\eta), \tag{32}$$

$$\mathcal{L}_2[g_m(\eta) - \chi_m g_{m-1}(\eta)] = \hbar_g R_m^g(\eta), \tag{33}$$

$$f_m(0) = f_m'(0) = g_m(0) = f_m'(\infty) = g_m(\infty) = 0, \tag{34}$$

where

$$R_m^f(\eta) = f_{m-1}''' + 2\sum_{k=0}^{m-1} f_k f_{m-1-k}'' - \sum_{k=0}^{m-1} f'_k f'_{m-1-k} + \sum_{k=0}^{m-1} g_k g_{m-1-k} - S\left(\frac{\eta}{2}f_{m-1}'' + f'_{m-1}\right), \tag{35}$$

$$R_m^g(\eta) = g_{m-1}'' - 2\sum_{k=0}^{m-1} g_k f'_{m-1-k} + 2\sum_{k=0}^{m-1} f'_k g'_{m-1-k} - S\left(\frac{\eta}{2}g'_{m-1} + g_{m-1}\right), \tag{36}$$

$$\chi_m = \begin{cases} 0, & m \leq 1, \\ 1, & m > 1, \end{cases} \tag{37}$$

in which $f_m(\eta)$ and $g_m(\eta)$ denote the special solutions of Equations (32) and (33) and the $C_i(i = 1-5)$ integral constants are calculated by the (34) boundary conditions. Equations (32) and (34) can be solved using Mathematica for $m = 1, 2, 3 \ldots$.

4. Convergence of the HAM Solution

The homotopy analysis method includes the regulating parameter h, which controls the region of convergence and HAM solution approximation. To ensure that the solutions converge within the admissible spectrum of auxiliary parameter values and hf and hg, $h-curves$ were sketched for 15th-order approximation. The $h-curves$ are plotted in Figures 2 and 3. The admissible ranges of values of hf and hg are $-1.5 \leq h_f < -0.3$ and $-1.5 \leq h_g < -0.3$, these ranges vary with the change in parameters.

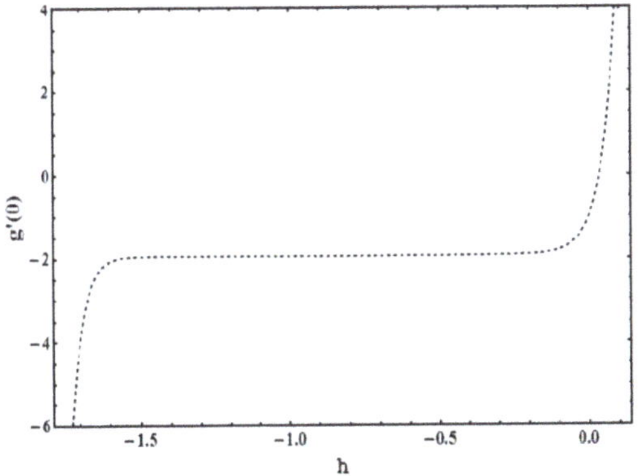

Figure 2. 15th-order $g'(0)$ for $a = 1$ and $S = -1$.

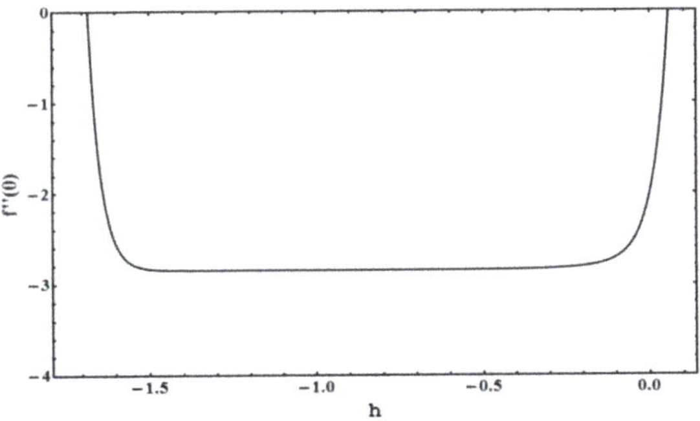

Figure 3. 15th-order $f''(0)$ for $a = 1$ and $S = -1$.

5. Results and Discussion

To solve Equations (8) and (9) homotopy analysis method (HAM) is applied as a subject to the boundary conditions (11). Homotopy analysis method is a strong analytical technique which is applied to obtain the convergent series solution of nonlinear differential equations. The convergence region for HAM through $h-curves$ are sketched and analyzed in Figures 3 and 4. Homotopy analysis method provides great freedom to obtain the convergent result. The convergence region varies for different values of a and S. Tables 1–4 represent the convergence of solution for different values of parameters. The error analysis of the obtained approximated results is as follows.

$$E_m = \int_0^{\infty} e_m^2(t)dt, \tag{38}$$

where $e_m(t)$ is the residual error of Equations (8) and (9) at the mth-order approximation. It is observed that 10th-order approximation is in good agreement with the numerical result.

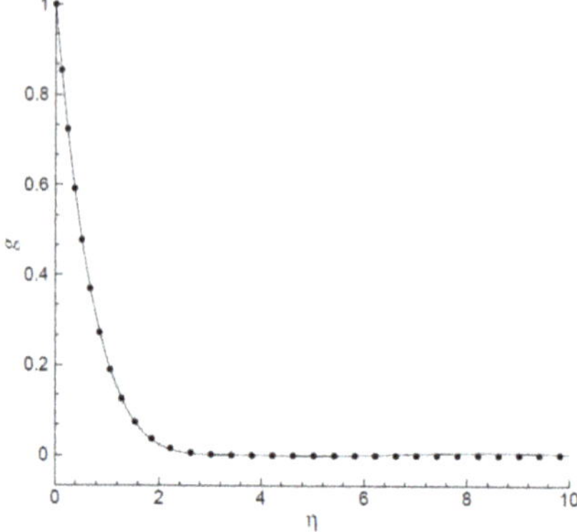

Figure 4. Comparison of convergence of $g(\eta)$ when $\hbar = -0.333$, $a = 1$ and $S = -1$ (line: 10th-order, dots: 5th-order).

Table 1. Comparison of the numerical result [5] with homotopy analysis method (HAM) convergent result when $a = 1$, $S = -1$ and $\hbar_f = -1/3$, $\hbar_g = -1/4$.

Order	$f''(0)$	$g'(0)$	*error f*	*error g*
2nd	−0.7673	−1.196	0.311	0.044
4th	−0.6945	−1.246	0.021	0.0067
6th	−0.6681	−1.264	0.0046	0.0021
8th	−0.6581	−1.270	0.00091	0.000715
10th	−0.6543	−1.271	0.00014	0.00023

Numerical result [5] $f'(0) = -0.6520, g'(0) = -1.2716$.

Table 2. Comparison of the numerical result [5] with HAM convergent result when $a = 1$, S = −1/10 and $\hbar_f = -1/3$, $\hbar_g = -1/4$.

Order	$f''(0)$	$g'(0)$	*error f*	*error g*
2nd	−0.9642	−1.3321	0.031	0.379
4th	−0.9374	−1.4162	0.0047	0.0055
6th	−0.9262	−1.446	0.00075	0.00091
8th	−0.9217	−1.458	0.00012	0.00016
10th	−0.9200	−1.4627	0.000018	0.000033

Numerical result [5] $f'(0) = -0.9189, g'(0) = -1.4656$.

Table 3. Comparison of the numerical result [5] with HAM convergent result when $a = 2$, S = −1/10 and $\hbar_g = \hbar_f = -1/5$.

Order	$f''(0)$	$g'(0)$	*error f*	*error g*
2nd	−2.779	−1.658	0.408	0.543
4th	−2.9729	−1.847	0.0876	0.136
6th	−3.044	−1.924	0.0234	0.0437
8th	−3.072	−1.953	0.0071	0.018
10th	−3.082	−1.958	0.0024	0.012

Numerical result [5] $f'(0) = -3.1178, g'(0) = -2.0530$.

Table 4. Comparison of the numerical result [5] with HAM convergent result when $a = 1, S = -1/2$ and $\hbar_g = \hbar_f = -1/4$.

Order	$f''(0)$	$g'(0)$	*error f*	*error g*
2nd	−0.9062	−1.2760	0.1051	0.0221
4th	−0.8592	−1.3424	0.0283	0.0037
6th	−0.8319	−1.3654	0.0077	0.00082
8th	−0.8172	−1.3741	0.0021	0.00021
10th	−0.8093	−1.3774	0.00058	0.00006

Numerical result [5] $f''(0) = -0.8007$, $g'(0) = -1.3797$.

The convergence control parameter plays an important role. In Tables 5 and 6, the effect of $\hbar$ on convergence is shown. Tables 5 and 6 show that the convergence of the solution depends strongly on $\hbar$. It can be seen easily that for one set of $\hbar$ the convergence is faster than the other.

Table 5. The Convergence analysis of $f''(0)$ for different $\hbar$ when $a = 1$, and $S = -1/2$.

Order	$\hbar f$=−1/4,$\hbar g$=−1/5	Err	$\hbar f$=−1/5,$\hbar g$=−1/4	Err
2nd	−0.9007	0.1062	−0.9243	0.1405
4th	−0.8535	0.0283	−0.8826	0.0514
6th	−0.8282	0.0076	−0.8658	0.0312
8th	−0.8149	0.0021	−0.8325	0.0069
10th	−0.8080	0.0005	−0.8201	0.0025

Numerical result [5] $f''(0) = -0.8007$.

Table 6. The Convergence analysis of $g'(0)$ for different $\hbar$ when $a = 1$, and $S = -1/2$.

Order	$\hbar f$=−1/4,$\hbar g$=−1/5	Err	$\hbar f$=−1/5,$\hbar g$=−1/4	Err
2nd	−1.2345	0.03718	−1.2786	0.022
4th	−1.3157	0.0087	−1.3445	0.0036
6th	−1.3502	0.0024	−1.3578	0.0016
8th	−1.366	0.0007	−1.3737	0.00021
10th	−1.3734	0.0002	−1.3768	0.00006

Numerical result [5] $g'(0) = 1.3797$.

In Figures 5 and 6, the comparison of 5th-order approximation with 10th-order approximation is shown, which again provide the facts for convergence. The Mathematica software is used to compute the results for higher-order approximation. As the given problem is highly nonlinear, the computation time increases if higher-order approximation is computed or increases the value of the parameters. For $a = 1$ and $S = 0$, the given problem becomes a special case as mentioned in the numerical paper [5]. Table 7 provides the convergence result for this special case as well. The results obtained in the present research for this special case are also in very good agreement with the numerical result. This shows the strength of homotopy analysis methods. It is found that for small S, $f''(0)$ decreases with the increase of 'a' as shown in Tables 2 and 3. Figure 7 represents the velocity distribution for different values of a. It is observed that with the increase in disk stretching parameter the velocity decreases. Figures 8 and 9 show that with the decrease in the unsteadiness parameter, both tangential and radial velocities increase.

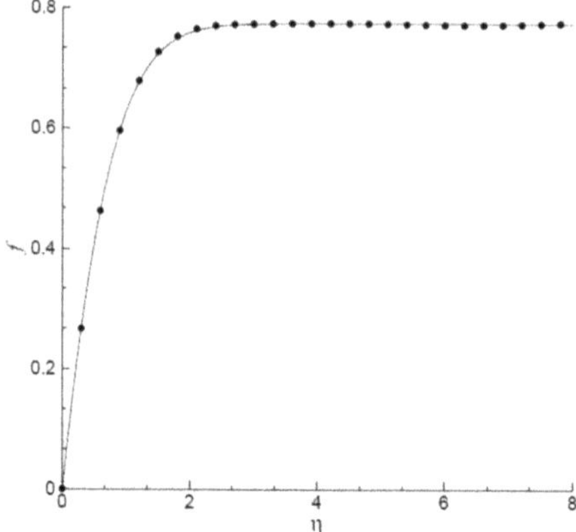

Figure 5. Comparison of convergence of $f(\eta)$ when $\hbar = -0.333$, $a = 1$ and $S = -1$ (line: 10th-order, dots: 5th-order).

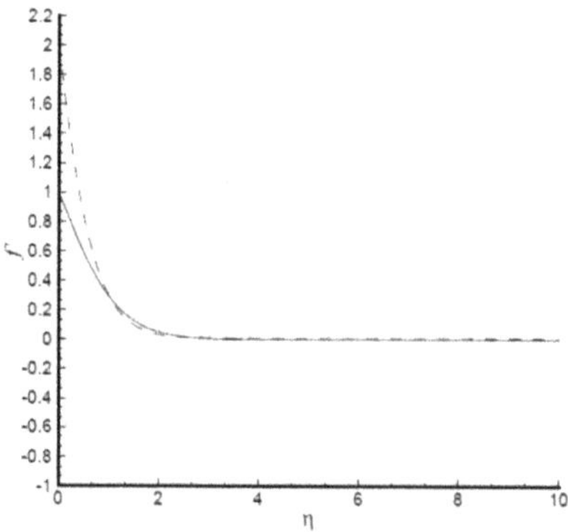

Figure 6. For $S = -1/2$ solid line: $a = 1$, Dashed line: $a = 2$ 10th order HAM approximation for $f'(\eta)$.

Table 7. Comparison of the numerical result [5] with HAM convergent result for special case when analysis = 1, $S = 0$, and $\hbar f = -28/100$, $\hbar g = -1/3$.

Order	f''	g'	error f	error g
5th	−1.1785	−1.44639	0.00044518	0.00019581
10th	−1.1751	−1.45359	0.00001437	8.13×10^{-7}
15th	−1.1739	−1.45402	3.86×10^{-7}	1.15×10^{-8}
20th	−1.1737	−1.45406	2.02×10^{-8}	6.42×10^{-10}

Numerical result [5] $f''(0) = -1.1737$, $g'(0) = -1.4541$.

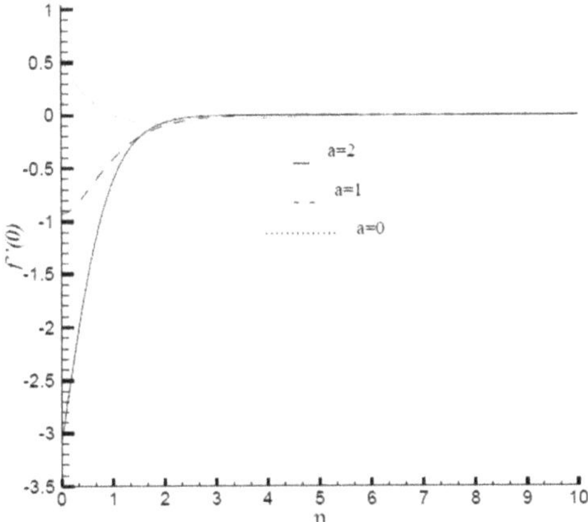

Figure 7. For $S = -1/10$ 10th-order HAM approximation for $f''(\eta)$.

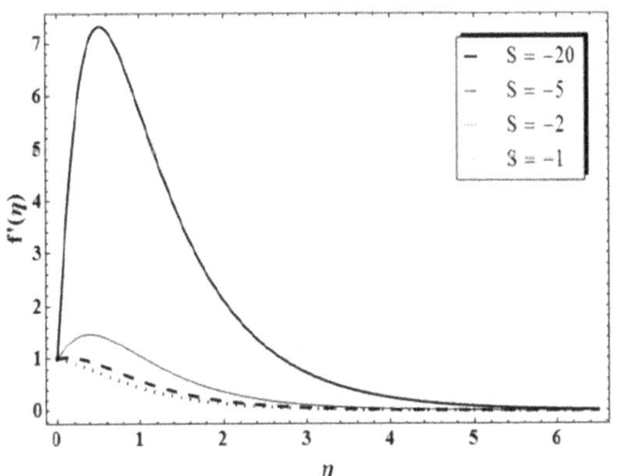

Figure 8. Variation of $f'(\eta)$ for different values of unsteadiness parameter for $a = 1$.

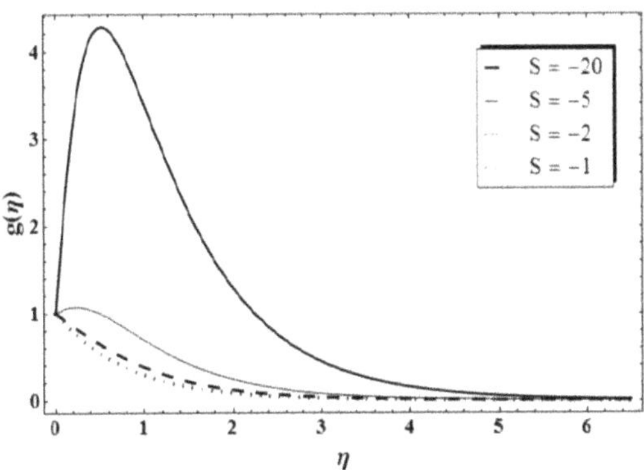

Figure 9. Variation of $g(\eta)$ for different values of unsteadiness parameter for $a = 1$.

6. Conclusions

In this research, viscous axisymmetric flow is studied on a stretchable rotating disk with deceleration. It is found that Navier–Stokes equation admits a similarity solution, which depends on non-dimensionalized parameters S and a measuring unsteadiness and disk stretching, respectively. The resulting group of nonlinear ordinary differential equations is then solved analytically using homotopy analysis method (HAM). In numerical paper [5] it is mentioned that there are two solution branches. The upper solution branch is physically feasible, but the lower solution branch may not be practically possible. Here, the author has discussed and evaluated the outcome for a physical solution from the upper field.

The main results are summarized as

- Results obtained by homotopy analysis method are in good agreement with existing numerical results;
- All the velocity profiles decrease with an increase in unsteadiness parameter S;
- Radial and axial velocity of the flow increases with the increase in disk stretching parameter a, whereas tangential velocity shows a decreasing trend with an increase in a;
- Variation trend decays with faster velocity to the ambient for fast deceleration as compared to the slow deceleration of the disk.

Funding: This research received funding from KFUPM.

Acknowledgments: The author wishes to express his thanks for the support received from King Fahd University of Petroleum and Minerals.

Conflicts of Interest: Author declare no conflicts of interest with any one.

References

1. Turkyilmazoglu, M. Three-dimensional MHD stagnation flow due to a stretchable rotating disk. *Int. J. Heat Mass Transf.* **2012**, *55*, 6959–6965. [CrossRef]
2. Von Kármán, T. Über Laminar Und Turbulente Reibung. *J. Appl. Math. Mech.* **1921**, *1*, 233–252.
3. Fang, T.; Zhang, J. Flow between two stretchable disks-An exact solution of the NavierStokes equations. *Int. Commun. Heat Mass Transf.* **2008**, *35*, 892–895. [CrossRef]
4. Rashidi, M.M.; Ali, M.; Freidoonimehr, N.; Nazari, F. Parametric analysis and optimization of entropy generation in unsteady MHD flow over a stretching rotating disk using articial neural network and particle swarm optimization algorithm. *Energy* **2013**, *55*, 1–14. [CrossRef]
5. Fang, T.; Hua, T. Unsteady viscous flow over a rotating stretchable disk with deceleration. *Commun. Nonlinear Sci. Numer. Simul.* **2012**, *17*, 5064–5072. [CrossRef]
6. Liao, S. On the homotopy analysis method for nonlinear problems. *Appl. Math. Comput.* **2004**, *147*, 499–513. [CrossRef]
7. Nadeem, S.; Awais, M. Thin film flow of an unsteady shrinking sheet through porous medium with variable viscosity. *Phys. Lett. A* **2008**, *372*, 4965–4972. [CrossRef]
8. Nadeem, S.; Ali, M. Analytical solutions for pipe flow of a fourth grade fluid with Reynold and Vogel's models of viscosities. *Commun. Nonlinear Sci. Numer. Simul.* **2009**, *14*, 2073–2090. [CrossRef]
9. Nadeem, S.; Abbasbandy, S.; Hussain, M. Series solutions of boundary layer flow of a Micropolar fluid near the stagnation point towards a shrinking sheet. *Z. Fur Nat.* **2009**, *64*, 575–582. [CrossRef]
10. Nadeem, S. Thin film flow of a third grade fluid with variable viscosity. *Z. Fur Nat.* **2009**, *64*, 553–558. [CrossRef]
11. Nadeem, S.; Hussain, A. MHD flow of a viscous fluid on a non-linear porous shrinking sheet by Homotopy analysis method. *Appl. Math. Mech.* **2009**, *30*, 1569–1578. [CrossRef]
12. Khan, H.; Ram, N.M.; Vajravelu, K.; Liao, S.J. The explicit Series Solution of SIR and SIS Epidemic Models. *Appl. Math. Comput.* **2009**, *215*, 653–669. [CrossRef]
13. Khan, H.; Liao, S.J.; Ram, N.M.; Vajravelu, K. An analytical solution for a nonlinear time delay model in Biology. *Commun. Nonlinear Sci. Numer. Simul.* **2009**, *14*, 3141–3148. [CrossRef]
14. Khan, H.; Xu, H. Series Solution of Thomas Fermi Atom Model. *Phys. Lett. A* **2007**, *365*, 111–115. [CrossRef]
15. Liao, S. A short review on the homotopy analysis method in fluid mechanics. *J. Hydrodyn.* **2010**, *22*, 882–884. [CrossRef]

Article

A Theoretical Analysis for Mixed Convection Flow of Maxwell Fluid between Two Infinite Isothermal Stretching Disks with Heat Source/Sink

Nargis Khan [1], Hossam A. Nabwey [2,3], Muhammad Sadiq Hashmi [4], Sami Ullah Khan [5] and Iskander Tlili [6,7,*]

1 Department of Mathematics, The Islamia University of Bahawalpur, Bahawalpur 63100, Pakistan; nargiskhan49@gmail.com
2 Department of Mathematics, College of Science and Humanities in Al-Kharj, Prince Sattam bin Abdulaziz University, Al-Kharj 11942, Saudi Arabia; eng_hossam21@yahoo.com
3 Department of Basic Engineering Science, Faculty of Engineering, Menoufia University, Shebin El-Kom 32511, Egypt
4 Department of Mathematics, The Govt. Sadiq College Women University, Bahawalpur 63100, Pakistan; mshashmi@gscwu.edu.pk
5 Department of Mathematics, COMSATS University Islamabad, Sahiwal 57000, Pakistan; sk_iiu@yahoo.com
6 Department for Management of Science and Technology Development, Ton Duc Thang University, Ho Chi Minh City 758307, Vietnam
7 Faculty of Applied Sciences, Ton Duc Thang University, Ho Chi Minh City 758307, Vietnam
* Correspondence: iskander.tlili@tdtu.edu.vn

Received: 10 December 2019; Accepted: 25 December 2019; Published: 27 December 2019

Abstract: The aim of this current contribution is to examine the rheological significance of Maxwell fluid configured between two isothermal stretching disks. The energy equation is also extended by evaluating the heat source and sink features. The governing partial differential equations (PDEs) are converted into the ordinary differential equations (ODEs) by using appropriate variables. An analytically-based technique is adopted to compute the series solution of the dimensionless flow problem. The convergence of this series solution is carefully ensured. The physical interpretation of important physical parameters like the Hartmann number, Prandtl number, Archimedes number, Eckert number, heat source/sink parameter and the activation energy parameter are presented for velocity, pressure and temperature profiles. The numerical values of different involved parameters for skin friction coefficient and local Nusselt number are expressed in tabular and graphical forms. Moreover, the significance of an important parameter, namely Frank-Kamenetskii, is presented both in tabular and graphical form. This particular study reveals that both axial and radial velocity components decrease by increasing the Frank–Kamenetskii number and stretching the ratio parameter. The pressure distribution is enhanced with an increasing Frank–Kamenetskii number and stretching ratio parameter. It is also observed that thetemperature distribution increases with the increasing Hartmann number, Eckert number and Archimedes number.

Keywords: maxwellfluid; mixed convection; isothermal stretching disks; homotopy analysis method

1. Introduction

The mixed convection flow is the combination of both coupled free and forced convection, and is a topic of particular interest from an engineering (aerospace and chemical engineering) point of view in the past few years. A diverse significance of such a phenomenon may appear in various electronic devices, nuclear reactors, food industries, energy storage, era of astrophysics, lubrication phenomenon, fire control, chemical metallurgical, etc. The phenomenon of free convection is resulted due to the temperature difference in fluid particles associated with isothermal stretching disks.

The involvement of magnetic force in the heat transfer processes between stretching disks is termed as forced convection. In a mixed convection flow, the Archimedes number represents the comparative contribution of natural to forced convection. It is a well justified fact that the phenomenon of free convection becomes more prevailing over forced convection when the Archimedes number is larger than unity. In the modern era of science, the flows caused by heat supplied in the presence of transport processes which occurred due to chemical reactions gained the attention of investigators due to numerous applications in several industrial processes. Arrhenius kinetics is adopted for modeling such reactions, where the flow is thermally obsessed by exothermic surface reaction. Maleque [1] studied the effects of exothermic/endothermic chemical reactions in the presence of energy activation over a porous flat plate. The impact of nonlinear thermal radiation and activation energy in the flow of Cross nanofluid has been reported by Khan et al. [2]. Shafique et al. [3] examined the flow of Maxwell fluid along with activation energy features in a rotating frame. A numerically-based continuation for viscous fluid flow in the presence of activation energy and slip factors has been pointed out by Awad et al. [4]. Another interesting contribution on the flow of viscoelastic fluid in presence of activation energy was investigated by Hsiao [5]. According to this study, the obtained observations can be used to enhance the manufacturing and thermal extrusion systems. The mixed convection flow on chemically reactive surfaces for external flow in the presence of porous medium was investigated by Merkin and Mahmood [6]. Similar studies were also performed by Minto et al. [7] for a vertical surface. We also acknowledge the interesting study presented by Chou and Tsern [8], in which they presented experimentally-based results regarding mixed convection flow in a horizontal channel with constant heat flux conditions.

In recent years, the stretched flows of electrically conducting materials under the influence of magnetic force have attained attention due to diverse engineering and medical applications. Some valuable applications of this phenomenon may include nuclear reactors, fission and fusion reactions, plasma, metallurgical processes, the exploration of oil, thermal conductors, magnetohydrodynamic (MHD) generators, etc. The MHD flow passing in arteries is important because of diverse physiological processes. For example, the flow of blood can be effectively controlled via an addition of the mixing of samples, heat transportation and interaction of the magnetic field. Many authors performed an extensive analysis regarding the MHD flow of various fluid models with different geometries. Nadeem et al. [9] investigated the impact of magnetic force in viscous nanofluid flow configured by a curved surface. Ahmed et al. [10] performed some numerical computations while explaining the thermophysical consequences in nanofluid flow subjected to magnetic force. Khan et al. [11] successfully obtained the dual solution for the combined heat and mass flow of magnetized nanoparticles over a curved surface. The oscillatory flow of micropolar nanofluid subjected to magnetic force has been numerically inspected by Sadiq et al. [12].

The study of non-Newtonian fluids is important due to their wide range of applications in engineering, physiology, the chemical and petroleum industries. The non-Newtonian fluid models capture a nonlinear relationship between shear stress and deformation rate in contrast to the viscous materials. The traditional examples of such fluids include paints, blood, paste, jell, apple source, etc. The non-Newtonian boundary layer flow due to stretching surfaces has been paid a great attention by scientists due to interesting industrial and engineering applications like glass fiber manufacturing, paper production, plastic films, crystal growing and in the processing of cooling bath of metallic sheets. In order to study the physical properties of non-Newtonian fluids, various models have been introduced in the literature. The classification of non-Newtonian models can be referred as rate type, differential type and integral type fluids. In the category of rate type, Maxwell fluid is considered as a subclass of rate type liquids which accomplishes the relaxation time features. The examples of Maxwell fluid include crude oil, toluene, polymer solution, etc. Haris [13] suggested the boundary layer equations for two-dimensional flow of Maxwell fluid. After that the analysis of boundary layer flow and heat transfer over a stretching surface by using the constitutive equation of Maxwell fluid was carried out by several researchers. For instance, Hayat et al. [14] discussed the series solution of upper-convected Maxwell fluid over a porous stretching plate.

The effects of thermal radiation on the MHD flow of a Maxwell fluid over a stretching surface were examined by Aliakbar et al. [15]. Two-dimensional stagnation-point flow of upper-convected Maxwell fluid (UCM) over a stretching sheet has been determined by Hayat et al. [16]. They used the homotopy analysis method (HAM) to solve the resulting nonlinear differential equations. Prasad et al. [17] discussed the effects of temperature-dependent viscosity, thermal conductivity and internal heat generation/absorption features in the MHD flow of upper-convected Maxwell fluid configured by a stretched surface. Khan et al. [18] examined the flow of Maxwell fluid in a channel with oscillating walls under the action of a magnetic field. The analysis for Maxwell fluid in the presence of a heat transfer phenomenon over coaxially rotating disks has been depicted by Ahmed et al. [19].

The fluid flow encountered by a rotating and stretching disk has gained serious importance in the last years due to a large number of physical applications for both physical and theoretical aspects. Some emerging applications of such flows includes a rotor-stator system, MHD generators, turbine engines, aircraft engines, spin coating, centrifugal pumps, flow-through swept wings, shrouded-disks rotation, rotating electrodes, centrifuges, hydraulic press, boilers, condensers, etc. Merkin and Chaudhary [20] investigated the flow of viscous fluid induced by stretching a disk in the presence of an exothermic surface reaction. Gorder et al. [21] reported the analytical solution for flow encountered by stretching disks. Khan et al. [22] studied the mixed convection flow induced by exothermal and isothermal stretching disks analytically.

In the present analysis, we study an incompressible mixed convection flow of Maxwell fluid between infinite isothermal stretching disks in the presence of heat absorption/generation, activation energy and chemical reaction features. In fact, this work is the extension of Gorder et al. [21] in three directions: Firstly, by considering Maxwell fluid, secondly by including activation energy consequences, and lastly by taking heat source and sink features. Considering the literature survey, it is noted that this present analysis has not been investigated yet and presented for the first time in literature. The study of the mixed convection flow of non-Newtonian fluid encountered enormous applications in nuclear engineering, chemical engineering and petroleum industries. The considered flow problem contained the impact of activation energy, which includes diverse industrial and engineering significance, like oil emulsion, food processing, chemical processes and geothermal reservoirs. The problem is solved analytically via the homotopy analysis method, and the results are discussed through pictorial and tabular representations.

2. Mathematical Formulation

In the current analysis, a non-Newtonian fluid is configured between two infinite stretching disks. It is assumed that flow is axisymmetric and steady. The rheological aspects of non-Newtonian material have been deliberated by using the famous Maxwell fluid model which occupies the space $0 < z < d$. The disks are separated distance d from each other as shown in Figure 1. The flow is generated due to the stretching of both disks in the radial direction. It is assumed that both (upper and lower) disks are isothermal in nature at temperatures T_1 and T_2, respectively. The analysis is performed by opting for cylindrical coordinates (r, θ, z). All the involved expressions are independent of θ due to axisymmetry. Following Merkin et al. [20], the expressions for first order non-isothermal reaction are represented in following form

$$A \rightarrow B + heat, \qquad rate = k_0 a_0 e^{-E/R_1 T}. \tag{1}$$

These above relations are known as Arrhenius kinetics, where E signifies the activation energy, B is a product species, R_1 is the gas constant, k_0 is the chemical reaction, a_0 the reactant concentration and T is the fluid temperature. The flow equations for the axisymmetric flow of Maxwell fluid can be expressed as [20–22]:

$$\frac{1}{r}\frac{\partial}{\partial r}(ru) + \frac{\partial w}{\partial z} = 0, \tag{2}$$

$$\begin{aligned} u\tfrac{\partial u}{\partial r} + w\tfrac{\partial u}{\partial z} &= -\tfrac{1}{\rho}\tfrac{\partial p}{\partial r} + \nu\Big(2\tfrac{\partial^2 u}{\partial r^2} + \tfrac{\partial^2 w}{\partial r \partial z} + \tfrac{\partial^2 u}{\partial z^2} + \tfrac{2}{r}\tfrac{\partial u}{\partial r} - 2\tfrac{u}{r^2}\Big) \\ &-\lambda_1\Big(w^2\tfrac{\partial^2 u}{\partial z^2} + 2uw\tfrac{\partial^2 u}{\partial r \partial z} + u^2\tfrac{\partial^2 u}{\partial r^2}\Big) + \tfrac{\delta B_0{}^2}{\rho}\Big(-u - \lambda_1 w\tfrac{\partial u}{\partial z}\Big) \\ &+g\beta\Big[(T - T_0) + \lambda_1\Big(u\tfrac{\partial T}{\partial r} + w\tfrac{\partial T}{\partial z} - \tfrac{\partial u}{\partial r}(T - T_0)\Big)\Big] \end{aligned} \tag{3}$$

$$u\frac{\partial w}{\partial r} + w\frac{\partial w}{\partial z} = -\frac{1}{\rho}\frac{\partial p}{\partial z} + \nu\left(\frac{\partial^2 w}{\partial r^2} + \frac{\partial^2 u}{\partial r \partial z} + 2\frac{\partial^2 w}{\partial z^2} + \frac{1}{r}\frac{\partial w}{\partial r} + \frac{1}{r}\frac{\partial u}{\partial r}\right) - \lambda_1\left(w^2\frac{\partial^2 u}{\partial z^2} + 2uw\frac{\partial^2 u}{\partial r \partial z} + u^2\frac{\partial^2 u}{\partial r^2}\right), \tag{4}$$

$$\begin{aligned} &\rho c_p\Big(u\tfrac{\partial T}{\partial r} + w\tfrac{\partial T}{\partial z}\Big) \\ &= 2\mu\Big(\Big(\tfrac{\partial u}{\partial r}\Big)^2 + \Big(\tfrac{\partial u}{\partial z}\Big)^2 + 2\Big(\tfrac{\partial w}{\partial r}\Big)\Big(\tfrac{\partial u}{\partial z}\Big) + \Big(\tfrac{u}{r}\Big)^2 + \Big(\tfrac{\partial w}{\partial r}\Big)^2 + \Big(\tfrac{\partial w}{\partial z}\Big)^2\Big) \\ &+ K_T\Big(\tfrac{\partial^2 T}{\partial r^2} + \tfrac{1}{r}\tfrac{\partial T}{\partial r} + \tfrac{\partial^2 T}{\partial z^2}\Big) - \tfrac{\delta B_0{}^2 u}{\rho} + Q_0(T - T_0) + Qk_0 a_0 e^{-E/R_1 T}, \end{aligned} \tag{5}$$

in which u and w are velocity components in the r and z directions, λ_1 is the relaxation time, p is the fluid pressure, ρ is the characteristic density, ν is the kinematic viscosity, K_T is the thermal conductivity of the fluid, T_0 the reference temperature given by $T_0 = \frac{T_1+T_2}{2}$, β denotes the thermal expansion coefficient, Q_0 denotes the heat generation/absorption coefficient while Q stands for the exothermicity factor. The imposed boundary conditions associated with the current flow problem are:

$$\begin{array}{lllll} u = ar, & w = 0, & p = \frac{a\mu\beta r^2}{4d^2}, & at & z = 0 \\ u = cr, & w = 0, & p = 0, & at & z = d, \\ T = T_1 \quad at & z = 0, & T = T_2, & at & z = d. \end{array} \tag{6}$$

In order to obtain the dimensionless form of above equations, we introduce the following similarity variables [21,22]:

$$\begin{array}{lll} u = arF(\eta), & w = adH(\eta), & \eta = \frac{z}{d} \\ u = arF(\eta), & w = adH(\eta), & \eta = \frac{z}{d} \\ p = a\mu\Big(P(\eta) + \frac{\beta_1 r^2}{4d^2}\Big), & & \theta(\eta) = \frac{E(T-T_0)}{R_1 T_0{}^2}. \end{array} \tag{7}$$

The above transformations lead to the following system:

$$\begin{aligned} H''' + \beta + \tfrac{R}{2}\Big(H'^2 &-2HH''\Big) \\ &= \lambda_1 aR\Big(H^2H''' - HH'H''\Big) - M(H' + \lambda_1 HH'') + A_r[2\theta \\ &+ \lambda_1 a(2H\theta' + H'\theta)], \end{aligned} \tag{8}$$

$$\theta'' - RPrH\theta' + \frac{EcPr}{\epsilon}\left(\frac{\delta^2}{2}(H'')^2 + 3H'^2\right) + \alpha\theta + K\exp\left(\frac{-1}{\epsilon(1+\epsilon\theta)}\right) = 0, \tag{9}$$

$$P' = H'' - RHH' - \lambda RH^2H'', \tag{10}$$

$$\left.\begin{array}{llll} H(0) = 0, & H(1) = 0, & H'(0) = -2, & H'(1) = -2\gamma, \\ \theta(0) = R_T, & \theta(1) = -R_T, & P(0) = 0. & \end{array}\right\} \tag{11}$$

where the stretching rate constant is γ, the Reynolds number R, the Hartmann number is M, Grashoff number G_r, heat source/sink parameter α, the Prandtl number Pr, Eckert number Ec, the Frank–Kamenetskii number K, constant temperature parameter R_T, activation energy parameter ϵ, the Archimedes number A_r and the dimensionless distance δ are defined as:

$$\left.\begin{array}{l} \gamma = \frac{c}{a}, R = \frac{ad^2}{v}, M = \frac{\sigma B_0{}^2}{a\rho}, G_r = \frac{g\beta T_0 r^3}{v^2}, \alpha = \frac{Q_0 d^2}{K_T}, Pr = \frac{\mu c_p}{K_T}, \\ Ec = \frac{ad^2}{C_p T_0}, K = \frac{Qk_0 a_0 d^2}{K_T \epsilon T_0}e^{-\frac{1}{\epsilon}}, R_T = \frac{T_1 - T_0}{\epsilon T_0}, \epsilon = \frac{R_1 T_0}{E}, A_r = \frac{G_r}{R^2}, \delta = \frac{r}{d} \end{array}\right\} \tag{12}$$

By differentiating Equation (8) with respect to similarity variable η, we have

$$\begin{aligned} H^{(iv)} - RHH''' &= \lambda R\left(HH'H''' + H^2H^{(iv)} - H'^2H'' - HH''^2\right) \\ &-MR\left[H'' + \lambda(H'H'' + HH''')\right] \\ &+\tfrac{RA_r\epsilon}{\delta^4}\left[2\theta' + \lambda(3H'\theta' + 2H\theta'' + H''\theta)\right]. \end{aligned} \tag{13}$$

where $\lambda = \lambda_1 a$, is the Deborah number. First of all, we solve Equation (12) subject to the boundary conditions (11) and then β can be evaluated by using Equation (8).

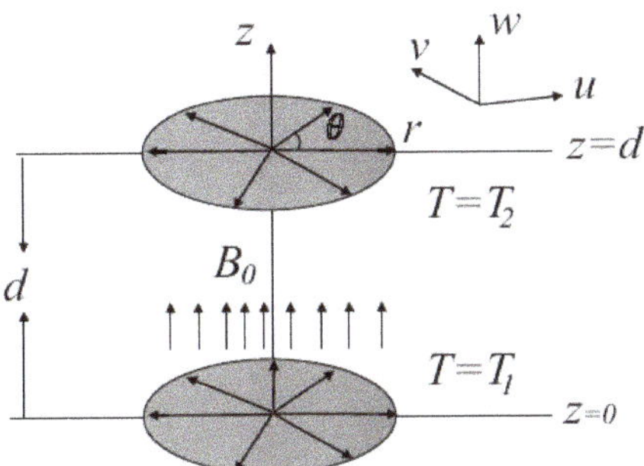

Figure 1. Geometry of the problem.

2.1. *Skin Friction Coefficient*

The expression for shear stress τ_w on the surface of the stretching disk is defined as [21,22]:

$$\tau_w = \tau_{rz}|_{z=0}, \tag{14}$$

The skin friction coefficients RC_{1f} and RC_{2f} at the lower and upper disks are:

$$C_{1f} = \frac{\tau_w}{\frac{1}{2}\rho(\delta r)^2} = \frac{\tau_{rz}|_{z=0}}{\frac{1}{2}\rho(\delta r)^2} = -R^{-1}H''(0), \tag{15}$$

$$C_{2f} = \frac{\tau_w}{\frac{1}{2}\rho(\delta r)^2} = \frac{\tau_{rz}|_{z=d}}{\frac{1}{2}\rho(\delta r)^2} = -R^{-1}H''(1). \tag{16}$$

2.2. *Local Nusselt Number*

The mathematical expressions for the local Nusselt number is represented as:

$$q_w = -\left(K_T\frac{\partial T}{\partial z}\right) = -\frac{K_T R_1 T_0^2}{Ed}\theta'(\eta), \tag{17}$$

The dimensionless form of this local Nusselt number at both (lower and upper) disks is [21,22]:

$$N_{1u} = \frac{dq_w}{K_T\epsilon T_0} = -\frac{dK_T\frac{\partial T}{\partial z}\big|_{z=0}}{K_T\epsilon T_0} = -\frac{dK_T R_1 T_0^2}{EdK_T\epsilon T_0}\theta'(0) = -\theta'(0), \tag{18}$$

$$N_{2u} = \frac{dq_w}{K_T\epsilon T_0} = -\frac{dK_T\frac{\partial T}{\partial z}\big|_{z=d}}{K_T\epsilon T_0} = -\frac{dK_T R_1 T_0^2}{EdK_T\epsilon T_0}\theta'(1) = -\theta'(1). \tag{19}$$

3. HomotopyAnalysis Method

In our modern era of scientific research, many physical and engineering problems are modeled in the form of highly nonlinear differential equations which always remain challenging for mathematicians to suggest the analytical or numerical solutions. Among different analytical techniques, thehomotopy analysis method is one which can be used to compute the analytic solution of such problems with excellent convergence. This technique is free of a complicated discretization procedure like numerical methods. This analytical technique is free of any small or large parameter constraints. This method was originally introduced by Liao [23], and later on many researchers used this method for their solutions of various problems [24–36]. The initial guesses for $H(\eta)$ and $\theta(\eta)$ are given by:

$$H_0(\eta) = 2\eta(1-\eta)((1+\gamma)\eta - 1), \tag{20}$$

$$\theta_0(\eta) = \eta, \tag{21}$$

Defining auxiliary linear operators

$$L_H[y] = \frac{d^4 y}{d\eta^4}, \tag{22}$$

$$L_\theta[y] = \frac{d^2 y}{d\eta^2}, \tag{23}$$

Satisfying

$$L_H\left[C_1 + C_2\eta + C_3\eta^2 + C_4\eta^3\right] = 0, \tag{24}$$

$$L_\theta[C_5 + C_6\eta] = 0, \tag{25}$$

where $C_i(i = 1-6)$ are constants.

4. Convergence of Obtained Solution

It is a well-established fact that the convergence rate of HAM solutions is strictly based on non-zero auxiliary parameters $\hbar_H$ and $\hbar_\theta$. The suitable selection of these parameters is quite useful for adjusting and controlling the obtained solution. The admissible range of these auxiliary parameters, the $\hbar$–curves for velocity and temperature distributions, is displayed in Figure 2a,b. These figures clearly demonstrate that the suitable values of $\hbar_H$ and $\hbar_\theta$ can be selected from $-2.0 \le \hbar_H \le -0.1$ and $-1.8 \le \hbar_\theta \le -0.3$. For present computations, the optimal values of $\hbar_H$ and $\hbar_\theta$ are taken $\hbar_H = -1$ and $\hbar_\theta = -1.08$. The accuracy of obtained solution against various values of emerging parameters is shown in Table 1. It is seen that the accuracy of the HAM solution is obtained at the twentieth order of approximations.

Table 1. The convergence analysis of the homotopic solution with $R = 5$, $\gamma = 0.5$, $M = 0.5$, $\lambda = 0.2$, $A_r = 2$, $\delta = 3$, $Ec = 1$, $Pr = 1$, $\alpha = 0.5$, $K = 0.01$, $R_T = 2$, $\epsilon = 0.5$, $\hbar_H = -1$ and $\hbar_\theta = -1.08$.

Order of Approximation	$H''(0)$	$\theta'(0)$
11	9.79594	−1.91738
14	9.79619	−1.91686
16	9.79643	−1.91634
18	9.79667	−1.91582
20	9.79717	−1.91461
25	9.79717	−1.91461
30	9.79717	−1.91461

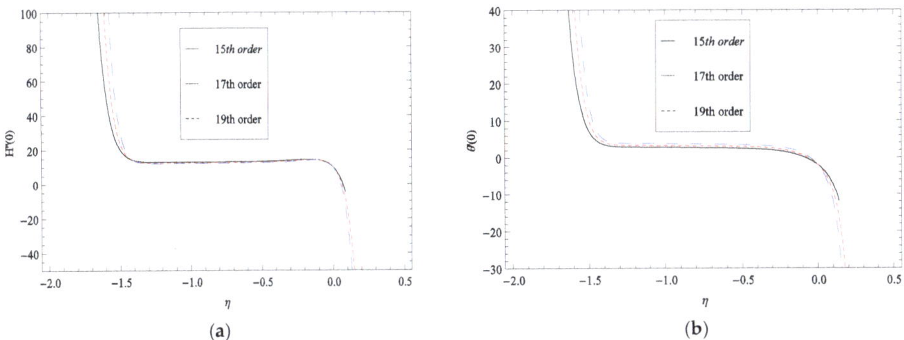

Figure 2. The h-curve for (**a**) velocity profile, (**b**) temperature profile when $R = 5,\ \gamma = 0.5,\ M = 0.5,$ $\lambda = 0.2,\ A_r = 2,\ \delta = 0.5,\ Ec = 1,\ Pr = 1,\ \alpha = 0.5,\ K = 0.001,\ R_T = 2$ and$\epsilon = 0.5$.

5. Validation of Solution

Before performing detailed graphical computations for flow parameters, we first compare our results with Gorder et al. [21] as a limiting case in Table 2. It is noted that an excellent accuracy of our results has been noted with these reported studies. Also Figure 3 shows the comparison of present results for the velocity profile computed via the homotopy analysis method for various values of the stretching ratio parameter with Gorder et al. [21]. It is found that present results have shown a convincible accuracy with Gorder et al. [21].

Table 2. Comparison of $H(\eta)$ and $H'(\eta)$ for different values of η with $\lambda = 0,\ R = 0$ and $M = 0$.

η	Gorder et al. [21]		Present Result (HAM)	
	$H(\eta)$	$H'(\eta)$	$H(\eta)$	$H'(\eta)$
0.0	0.000	−2.00	0.000	−2.00
0.2	−0.224	−0.360	−0.224	−0.360
0.4	−0.192	0.560	−0.192	0.560
0.6	−0.048	0.760	−0.048	0.760
0.8	0.064	0.240	0.064	0.240
1.0	0.000	−1.00	0.000	−1.000

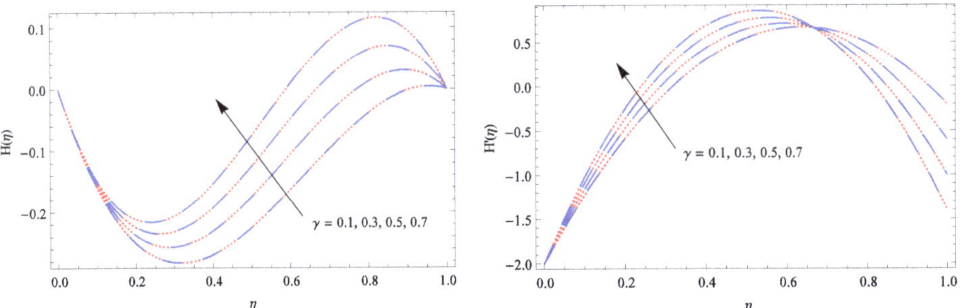

Figure 3. Comparison of velocity components $H(\eta)$ and $H'(\eta)$ for different values of stretching parameter γ, dotted red lines represents the HAM solution while blue lines denotes the solution by Gorder et al. [21].

6. Results and Discussion

The formulated ordinary differential equations are targeted analytically via the homotopy analysis scheme. The aim of this section is to examine the physical significance of each physical parameter on velocity, pressure and temperature distributions.

6.1. Velocity Distribution

Figures 4–6 are plotted to capture the influence of various parameters like the stretching ratio γ, Deborah number λ, Reynolds number R, Prandtl number Pr, heat source/sink parameter α, constant temperature parameter R_T, the Eckert number Ec, activation energy parameter ϵ, Frank–Kamenetskii number K, Hartmann number M and the Archimedes number A_r on the $r-$velocity component $H'(\eta)$ and $z-$direction velocity component $H(\eta)$. Figure 4a–f presents the effect of the Deborah number λ, stretching ratio γ, Eckert number Ec, Prandtl number Pr, dimensionless distance parameter δ and the Frank–Kamenetskii number K on the$r$- and z-components of velocity. Figure 3a prescribed the outcomes of the Deborah number λ on the$z$ component of velocity. It is observed that the r-component of velocity increases up to a certain range and later on decreases slightly. It can be justified physically, as the Deborah number is directly proportional to the relaxation time. In fact, it is the associated with the fluid relaxation time to the observation time. The smaller values of the Deborah number represent the viscous nature of fluid while a material having a higher Deborah number represents the solid nature of fluid. The effects of the stretching ratio γ, Eckert number Ec, dimensionless distance δ and the Frank–Kamenetskii number K on $H(\eta)$ and $H'(\eta)$ has been expressed in Figure 4b–e. From all these figures, it is noted that both $H(\eta)$ and $H'(\eta)$ are enhanced by varying these parameters. Figure 4f manifested the influence of the Hartmann number on the z component of velocity. A decay in the z component of velocity is observed for intensifying values of the Hartmann number. Physically, the larger values of this Hartmann number attributed strong drag force which resists the amplitude of flow.Figure 4g–h determined the effects of the Archimedes number A_r and Reynolds number R on axial and radial velocities. A retarded distribution of both components has been resulted with the variation of all these parameters. Since the Reynolds number represents the ratio of inertial force to viscous, therefore higher values of R become associated with larger inertial force which decay the velocity distribution effectively.

6.2. Pressure Distribution

The variation in pressure distribution $P(\eta)$ for various values of the stretching ratio parameter γ, dimensionless distance δ, Reynolds number R, Archimedes number A_r, constant temperature parameter R_T, activation energy parameter ϵ, Deborah number λ, Hartmann number M and the Frank–Kamenetskii number K is discussed in Figure 5a–h. Figure 5a,b show the change in $P(\eta)$ for diverse values of γ and δ. It is noted that pressure is an increasing function of γ and δ up to a specific height, and later on decreases gradually. However, a decreasing trend has been observed for maximum values of the Reynolds number R and the Archimedes number A_r (Figure 5c,d). The graphical explanation for the Deborah number λ and the Hartmann number M is presented in Figure 5e,f. The Deborah number specified the relaxation time to the observation time ratio which means that maximum values of λ correspond to larger relaxation time due to which the pressure distribution declined. Similarly, a decreasing trend in the pressure distribution is due to the fact that the Hartmann number is associated with Lorentz force, which efficiently controls the pressure distribution in the whole domain. Figure 5g determines the influence of the Frank–Kamenetskii number K on pressure distribution $P(\eta)$. A retarded pressure distribution has been examined with the variation of K. With the increase of K, the pressure distribution decreases up to maximum level.

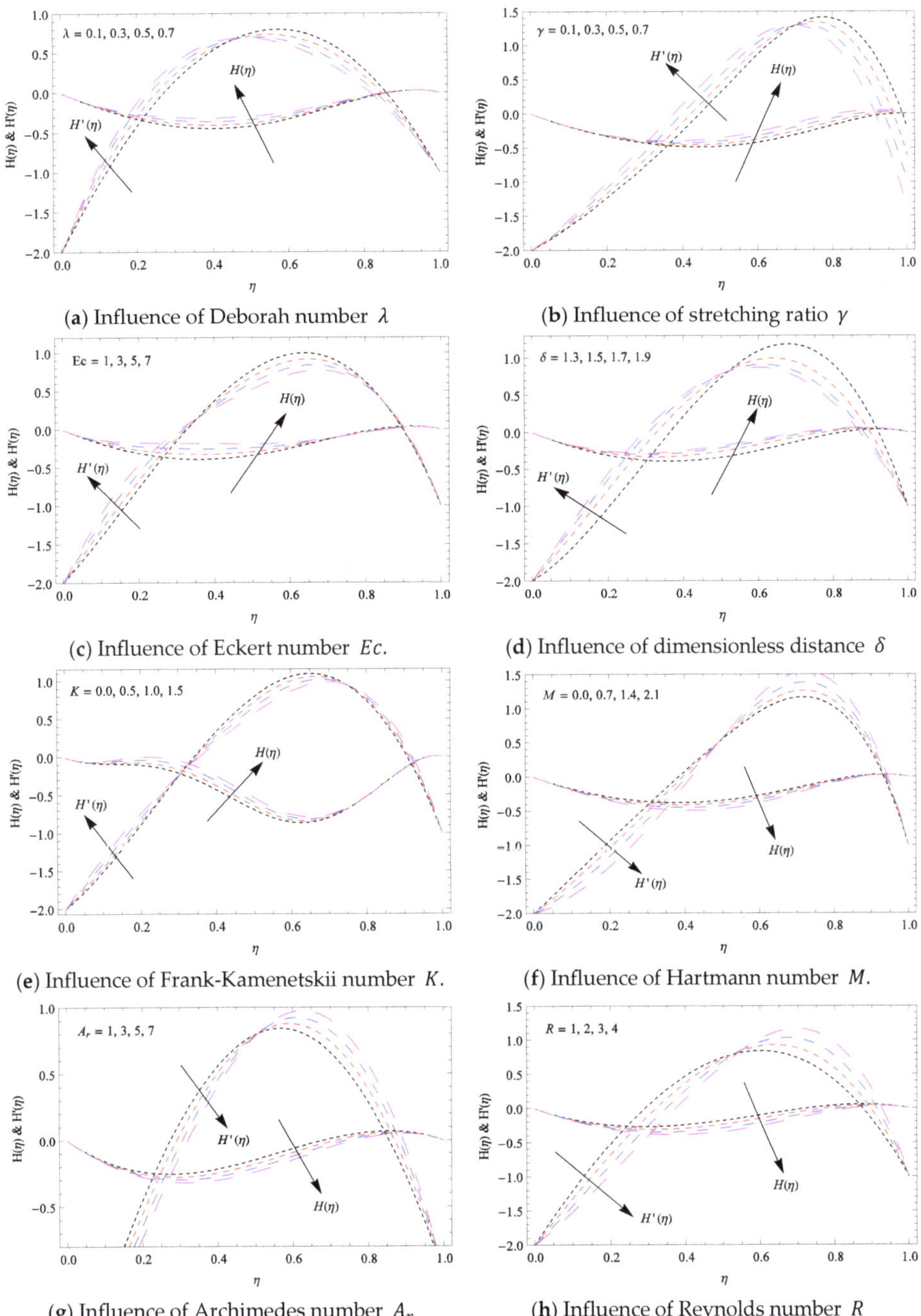

(**a**) Influence of Deborah number λ (**b**) Influence of stretching ratio γ

(**c**) Influence of Eckert number Ec. (**d**) Influence of dimensionless distance δ

(**e**) Influence of Frank-Kamenetskii number K. (**f**) Influence of Hartmann number M.

(**g**) Influence of Archimedes number A_r (**h**) Influence of Reynolds number R

Figure 4. In r and z components of velocities when $\gamma = 0.5,\ \hbar_H = -1,\ M = 0.5,\ \lambda = 0.2,\ Ec = 1,\ Pr = 1,$ $\alpha = 0.5,\ R = 5,\ \delta = 1.5,\ \epsilon = 0.5,\ A_r = 50,\ K = 0.5$ and $R_T = 1$.

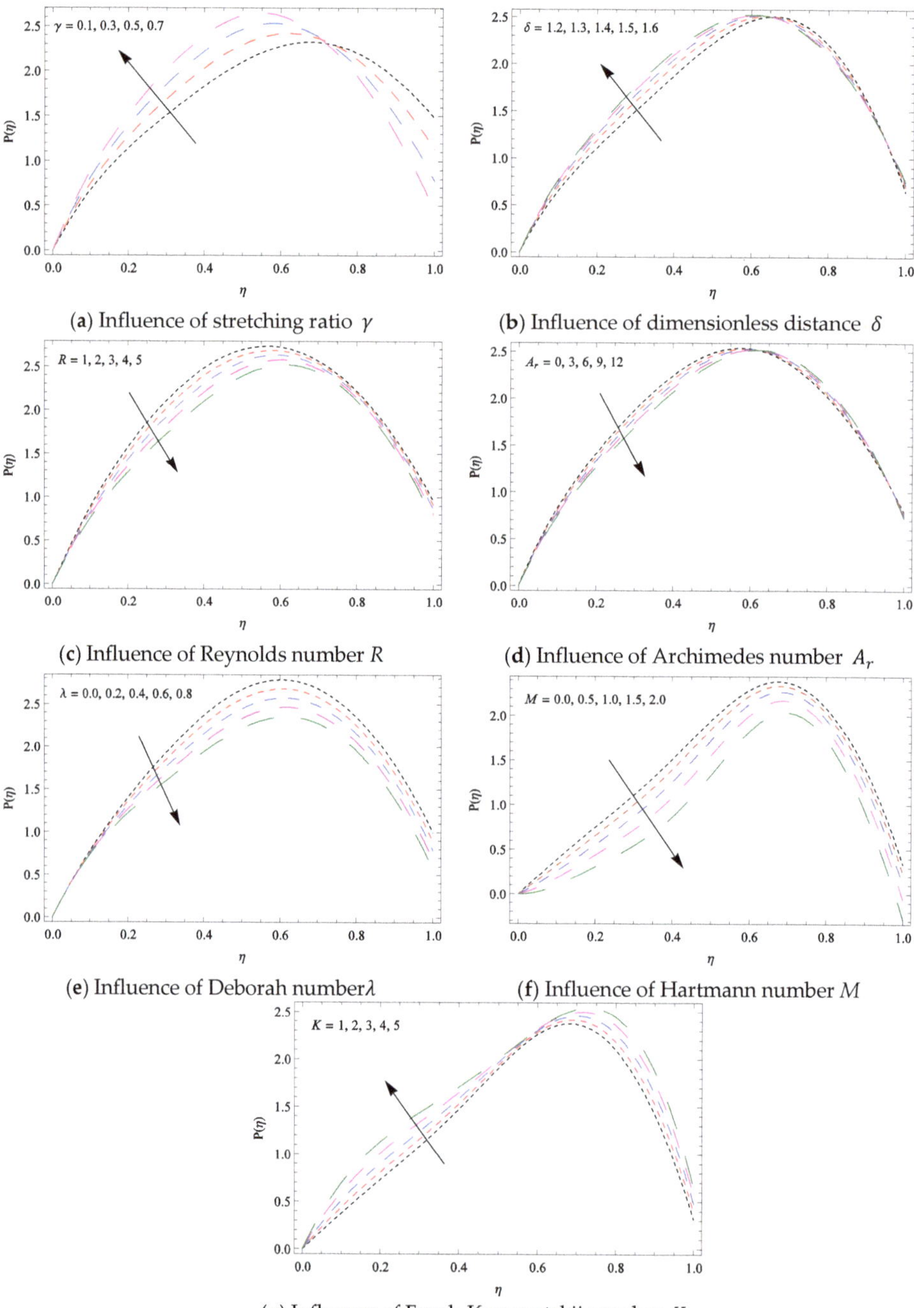

(**a**) Influence of stretching ratio γ (**b**) Influence of dimensionless distance δ

(**c**) Influence of Reynolds number R (**d**) Influence of Archimedes number A_r

(**e**) Influence of Deborah numberλ (**f**) Influence of Hartmann number M

(**g**) Influence of Frank-Kamenetskii number K.

Figure 5. Pressure distribution for different parameters with $\gamma = 0.5$, $\hbar_H = -1$, $M = 0.5$, $\lambda = 0.5$, $Ec = 1$, $Pr = 1$, $\alpha = 0.9$, $\delta = 1.5$, $R = 5$, $\epsilon = 0.5$, $A_r = 2$, $K = 0.5$ and $R_T = 2$.

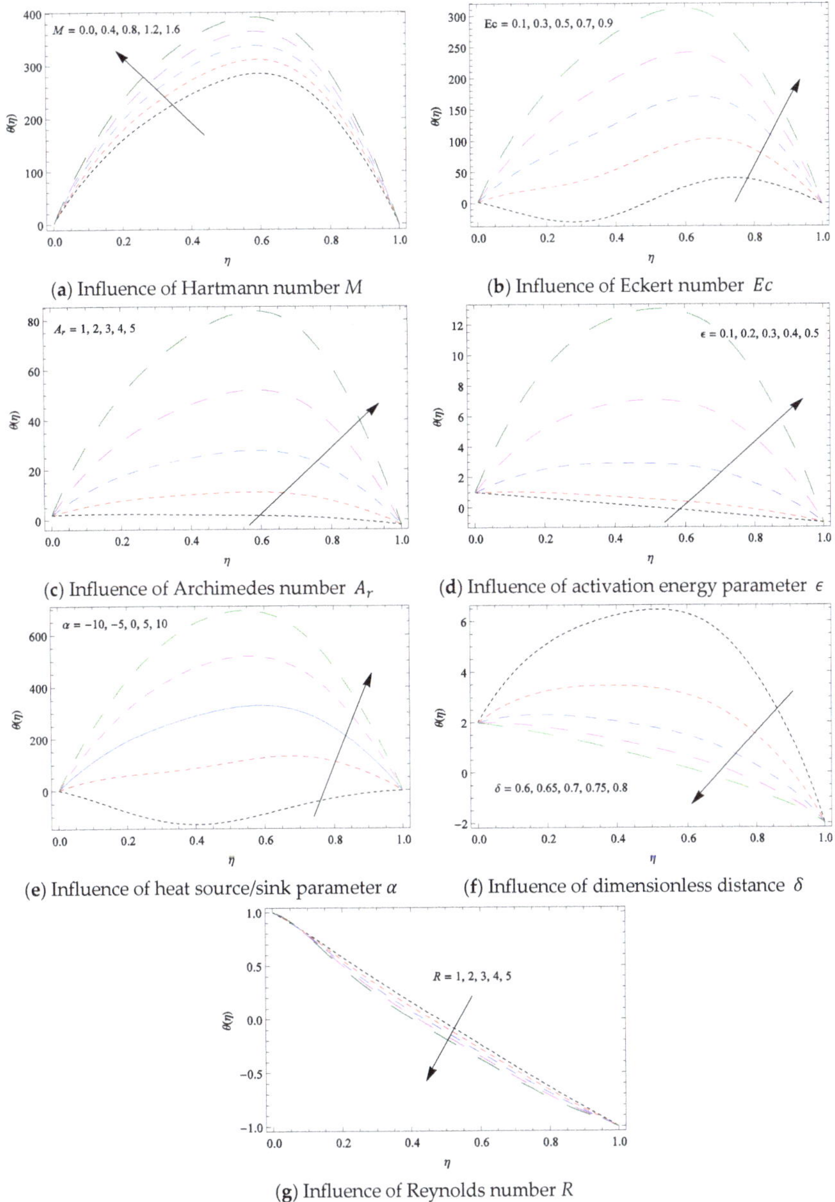

(**a**) Influence of Hartmann number M

(**b**) Influence of Eckert number Ec

(**c**) Influence of Archimedes number A_r

(**d**) Influence of activation energy parameter ϵ

(**e**) Influence of heat source/sink parameter α

(**f**) Influence of dimensionless distance δ

(**g**) Influence of Reynolds number R

Figure 6. Temperature profile for $\gamma = 0.5$, $\hbar_H = -1$, $\hbar_\theta = -1.08$, $M = 1$, $\lambda = 0.2$, $Ec = 1$, $R = 5$, $Pr = 1$, $\alpha = 0.5$, $\delta = 0.5$, $\epsilon = 0.5$, $A_r = 5$, $K = 0.01$ and $R_T = 1$.

6.3. Temperature Distribution

In order to examine the impact of the Hartmann number M, heat source/sink parameter α, Eckert number Ec, stretching ratio parameter γ, Archimedes number A_r, dimensionless distance δ and the Reynolds number R on temperature distribution $\theta(\eta)$, Figure 6a–g are prepared. Figure 6a captured the consequences of the Hartmann number M on temperature distribution $\theta(\eta)$. As expected, an enhanced temperature distribution is observed for larger values of M due to the interaction of Lorentz force. From Figure 6b, again an increment in temperature distribution has been noted for maximum values of the Eckert number *Ec*. The physical consequences of such trend may be attributed as heat due to viscous dissipation of fluid enhanced, due to which results an increment in $\theta(\eta)$. Figure 6c,d portrayed the impact Archimedes number A_r and activation energy parameter ϵ on $\theta(\eta)$. It is seen that temperature distribution enlarges with increasing both parameters. The activation energy plays a significant role in enhancement of many reaction processes. Figure 6e reports the influence of the heat source/sink constant on $\theta(\eta)$. It is noted that $\theta(\eta)$ increases in the case of heat source case $(\alpha > 0)$, while the opposite trend is noted for the heat sink case $(\alpha < 0)$. The physical aspect of such a trend may attribute, as in the case of heat source, more heat is added to the system, due to which the temperature distribution improved. On the contrary, due to the heat sink, heat is removed from the whole system which turns down the temperature distribution efficiently. From Figure 6f,g, a declining temperature distribution has been observed with maximum variation of dimensionless distance δ and Reynolds number *R*.

6.4. Physical Quantities of Interests

Figure 7a–c show the effect of different parameters like the stretching ratio parameter γ, Deborah number λ and Hartmann number *M* on the skin friction coefficient at upper and lower disks. From Figure 7a, a decreasing variation in this skin friction coefficient is examined with increasing γ. On contrary, the skin friction coefficient at both level disks is increased for maximum values of the Deborah number λ (Figure 7b). Figure 7c reveals that the wall shear stress for different values of the Hartmann number *M* is maximum at the upper level of the disk as compared to the lower level.

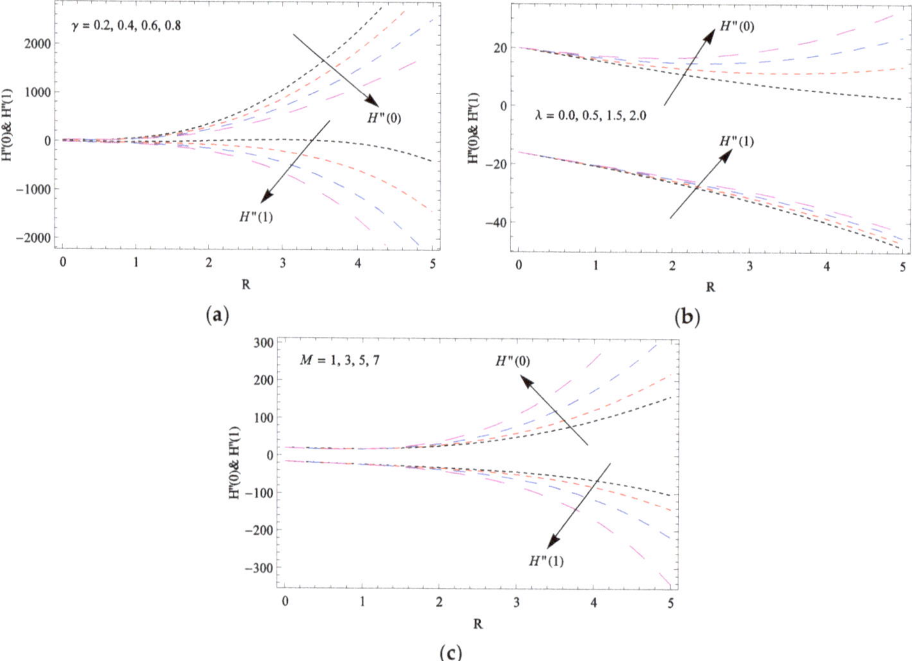

Figure 7. **(a–c)** The in skin friction coefficients at both disks for various values of γ, λ and M with $\gamma = 0.5, \hbar_H = -1, \hbar_\theta = -1.08, M = 1, \lambda = 2, Ec = 1, Pr = 1, \alpha = 0.5, \delta = 1.2, \epsilon = 0.5, A_r = 50$, $K = 0.01$ and $R_T = 1$.

Figure 8a,b show the effects of Hartmann number M and Eckert number Ec on the local Nusselt number at lower and upper disks. The variation in local Nusselt number at the upper disk is larger for both parameters.

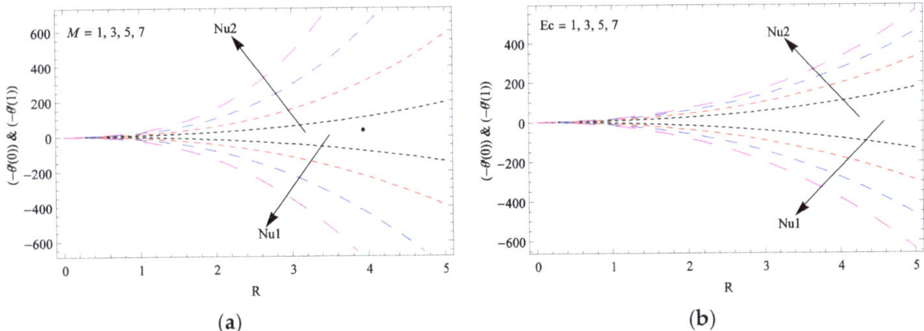

Figure 8. (**a**,**b**) The variation in Nusselt number at both disks for various values of M and Ec when $\gamma = 0.5$, $\hbar_H = -1$, $\hbar_\theta = -1.08$, $M = 1$, $\lambda = 2$, $Ec = 1$, $Pr = 1$, $\alpha = 0.5$, $\delta = 1.2$, $\epsilon = 0.5$, $A_r = 50$, $K = 0.01$ and $R_T = 1$.

The numerical iteration in the wall shear stress at the upper level of the disk RC_{1f} and lower level RC_{2f} are discussed in Table 3. The wall shear stress gets minimum values for stretching rate constant γ, Reynolds number R, and Hartmann number M. It is noted that rate of wall shear stress is relatively slower at the lower portion of the disk for all parameters. The variation for various parameters on the local Nusselt number is portrayed in Table 4. Again, the continuations are performed at both surfaces (upper surface N_{1u} and lower surface N_{2u}). This physical quantity increases with Pr and Ec. Finally, the numerical values of Frank–Kamenetskii against different values of γ, M, λ, R, Pr, Ec, δ, α, ϵ, A_r and R_T is shown in Table 5. The variation in Frank–Kamenetskii constant is slower for λ and A_r.

Table 3. Numerical variation in wall shear stress at both surfaces of moving disk.

γ	M	R	Pr	Ec	λ	δ	α	ϵ	A_r	R_T	K	Lower Disk	Upper Disk
0.2	01	05	1.0	1.0	0.2	0.5	0.5	0.5	02	2.0	0.1	−116.204	−252.335
0.4	01	05	1.0	1.0	0.2	0.5	0.5	0.5	02	2.0	0.1	−129.404	−262.989
0.6	01	05	1.0	1.0	0.2	0.5	0.5	0.5	02	2.0	0.1	−142.249	−272.563
0.5	1.0	05	1.0	1.0	0.2	0.5	0.5	0.5	02	2.0	0.1	−135.869	−267.915
0.5	1.5	05	1.0	1.0	0.2	0.5	0.5	0.5	02	2.0	0.1	−149.413	−273.824
0.5	2.0	05	1.0	1.0	0.2	0.5	0.5	0.5	02	2.0	0.1	−162.905	−281.867
0.5	01	1.0	1.0	1.0	0.2	0.5	0.5	0.5	02	2.0	0.1	−1.63619	−38.7856
0.5	01	2.0	1.0	1.0	0.2	0.5	0.5	0.5	02	2.0	0.1	−22.2454	−67.5845
0.5	01	3.0	1.0	1.0	0.2	0.5	0.5	0.5	02	2.0	0.1	−44.9617	−108.970
0.5	01	05	1.0	1.0	0.2	0.5	0.5	0.5	02	2.0	0.1	−135.869	−267.915
0.5	01	05	2.0	1.0	0.2	0.5	0.5	0.5	02	2.0	0.1	−102.816	−437.258
0.5	01	05	3.0	1.0	0.2	0.5	0.5	0.5	02	2.0	0.1	−69.7624	−606.601
0.5	01	05	1.0	1.0	0.2	0.5	0.5	0.5	02	2.0	0.1	−135.869	−267.915
0.5	01	05	1.0	1.5	0.2	0.5	0.5	0.5	02	2.0	0.1	115.683	−430.368
0.5	01	05	1.0	2.0	0.2	0.5	0.5	0.5	02	2.0	0.1	367.236	−592.821
0.5	01	05	1.0	1.0	0.2	0.5	0.5	0.5	02	2.0	0.1	−135.869	−267.915
0.5	01	05	1.0	1.0	0.4	0.5	0.5	0.5	02	2.0	0.1	−195.891	−369.982
0.5	01	05	1.0	1.0	0.6	0.5	0.5	0.5	02	2.0	0.1	−255.476	−472.018
0.5	01	05	1.0	1.0	0.2	0.5	0.5	0.5	02	2.0	0.1	−135.869	−267.915
0.5	01	05	1.0	1.0	0.2	0.6	0.5	0.5	02	2.0	0.1	−62.2384	−77.8090
0.5	01	05	1.0	1.0	0.2	0.7	0.5	0.5	02	2.0	0.1	−23.2608	−44.7235
0.5	01	05	1.0	1.0	0.2	0.5	0.1	0.5	02	2.0	0.1	−130.344	−270.583
0.5	01	05	1.0	1.0	0.2	0.5	0.3	0.5	02	2.0	0.1	−133.108	−269.244
0.5	01	05	1.0	1.0	0.2	0.5	0.5	0.5	02	2.0	0.1	−135.869	−267.915

Table 3. *Cont.*

γ	M	R	Pr	Ec	λ	δ	α	ϵ	A_r	R_T	K	Lower Disk	Upper Disk
0.5	01	05	1.0	1.0	0.2	0.5	0.5	0.1	02	2.0	0.1	−14.4518	−37.1367
0.5	01	05	1.0	1.0	0.2	0.5	0.5	0.3	02	2.0	0.1	−52.1849	−61.3395
0.5	01	05	1.0	1.0	0.2	0.5	0.5	0.5	02	2.0	0.1	−89.1381	−99.7870
0.5	01	05	1.0	1.0	0.2	0.5	0.5	0.5	3.5	2.0	0.1	24.7854	−1205.14
0.5	01	05	1.0	1.0	0.2	0.5	0.5	0.5	4.0	2.0	0.1	195.575	−1786.25
0.5	01	05	1.0	1.0	0.2	0.5	0.5	0.5	4.5	2.0	0.1	445.195	−2541.17
0.5	01	05	1.0	1.0	0.2	0.5	0.5	0.5	02	2.0	0.1	−135.869	−267.915
0.5	01	05	1.0	1.0	0.2	0.5	0.5	0.5	02	2.3	0.1	−185.374	−345.966
0.5	01	05	1.0	1.0	0.2	0.5	0.5	0.5	02	2.6	0.1	−247.552	−440.297
0.5	01	05	1.0	1.0	0.2	0.5	0.5	0.5	02	2.0	0.1	−155.783	−257.838
0.5	01	05	1.0	1.0	0.2	0.5	0.5	0.5	02	2.0	0.2	−154.907	−257.367
0.5	01	05	1.0	1.0	0.2	0.5	0.5	0.5	02	2.0	0.3	−154.003	−256.903

Table 4. Numerical variation in local Nusselt number at both surfaces of moving disk.

γ	M	R	Pr	Ec	λ	δ	α	ϵ	A_r	R_T	K	Lower Disk	Upper Disk
0.2	01	05	1.0	1.0	0.2	0.5	0.5	0.5	02	2.0	0.1	−43.2173	62.2902
0.4	01	05	1.0	1.0	0.2	0.5	0.5	0.5	02	2.0	0.1	−43.1456	64.2379
0.6	01	05	1.0	1.0	0.2	0.5	0.5	0.5	02	2.0	0.1	−43.1225	66.2978
0.5	1.0	05	1.0	1.0	0.2	0.5	0.5	0.5	02	2.0	0.1	−43.1714	65.2526
0.5	1.5	05	1.0	1.0	0.2	0.5	0.5	0.5	02	2.0	0.1	−49.2164	71.3174
0.5	2.0	05	1.0	1.0	0.2	0.5	0.5	0.5	02	2.0	0.1	−55.2913	77.3855
0.5	01	2.0	1.0	1.0	0.2	0.5	0.5	0.5	02	2.0	0.1	−1.93260	10.9702
0.5	01	3.0	1.0	1.0	0.2	0.5	0.5	0.5	02	2.0	0.1	−9.76591	20.5937
0.5	01	4.0	1.0	1.0	0.2	0.5	0.5	0.5	02	2.0	0.1	−20.8434	35.8615
0.5	01	05	1.0	1.0	0.2	0.5	0.5	0.5	02	2.0	0.1	−34.5174	58.1590
0.5	01	05	2.0	1.0	0.2	0.5	0.5	0.5	02	2.0	0.1	−59.2591	126.626
0.5	01	05	3.0	1.0	0.2	0.5	0.5	0.5	02	2.0	0.1	−70.4176	209.251
0.5	01	05	1.0	1.0	0.2	0.5	0.5	0.5	02	2.0	0.1	−34.5174	58.1590
0.5	01	05	1.0	1.5	0.2	0.5	0.5	0.5	02	2.0	0.1	−65.0929	88.3373
0.5	01	05	1.0	2.0	0.2	0.5	0.5	0.5	02	2.0	0.1	−95.5317	118.241
0.5	01	05	1.0	1.0	0.2	0.5	0.5	0.5	02	2.0	0.1	−37.1562	59.1911
0.5	01	05	1.0	1.0	0.4	0.5	0.5	0.5	02	2.0	0.1	−35.3967	58.5028
0.5	01	05	1.0	1.0	0.6	0.5	0.5	0.5	02	2.0	0.1	−33.6384	57.8153
0.5	01	05	1.0	1.0	0.2	0.5	0.5	0.5	02	2.0	0.1	−34.5174	58.1590
0.5	01	05	1.0	1.0	0.2	0.6	0.5	0.5	02	2.0	0.1	−4.57816	17.6285
0.5	01	05	1.0	1.0	0.2	0.7	0.5	0.5	02	2.0	0.1	−2.16759	8.23666
0.5	01	05	1.0	1.0	0.2	0.5	0.1	0.5	02	2.0	0.1	−32.5620	56.3592
0.5	01	05	1.0	1.0	0.2	0.5	0.3	0.5	02	2.0	0.1	−33.5476	57.2673
0.5	01	05	1.0	1.0	0.2	0.5	0.5	0.5	02	2.0	0.1	−34.5174	58.1590
0.5	01	05	1.0	1.0	0.2	0.5	0.5	0.1	02	1.0	0.1	−4.03154	4.78376
0.5	01	05	1.0	1.0	0.2	0.5	0.5	0.3	02	1.0	0.1	−7.31547	21.5100
0.5	01	05	1.0	1.0	0.2	0.5	0.5	0.5	02	1.0	0.1	−34.5174	58.1590
0.5	01	05	1.0	1.0	0.2	0.5	0.5	0.5	3.5	1.0	0.1	−128.255	180.574
0.5	01	05	1.0	1.0	0.2	0.5	0.5	0.5	4.0	1.0	0.1	−171.887	236.943
0.5	01	05	1.0	1.0	0.2	0.5	0.5	0.5	4.5	1.0	0.1	−221.713	301.095
0.5	01	05	1.0	1.0	0.2	0.5	0.5	0.5	02	2.0	0.1	−34.5174	58.1590
0.5	01	05	1.0	1.0	0.2	0.5	0.5	0.5	02	2.3	0.1	−43.9555	76.6222
0.5	01	05	1.0	1.0	0.2	0.5	0.5	0.5	02	2.6	0.1	−53.7965	97.8575
0.5	01	05	1.0	1.0	0.2	0.5	0.5	0.5	02	1.0	0.1	−35.1337	58.7998
0.5	01	05	1.0	1.0	0.2	0.5	0.5	0.5	02	1.0	0.2	−35.8183	59.5099
0.5	01	05	1.0	1.0	0.2	0.5	0.5	0.5	02	1.0	0.3	−36.5027	60.2180

Table 5. The numerical values of Critical Frank–Kamenetskii number for various parameters.

γ	M	R	Pr	Ec	δ	α	ϵ	A_r	R_T	λ	K_c
0.3	01	05	01	01	0.5	0.5	0.01	02	2.0	1.2	−10.8144
0.5	01	05	01	01	0.5	0.5	0.01	02	2.0	1.2	−10.8151
0.7	01	05	01	01	0.5	0.5	0.01	02	2.0	1.2	−10.8170
0.5	01	05	01	01	0.5	0.5	0.01	02	2.0	1.2	−10.8151
0.5	03	05	01	01	0.5	0.5	0.01	02	2.0	1.2	−10.7897
0.5	05	05	01	01	0.5	0.5	0.01	02	2.0	1.2	−10.7400
0.5	01	01	01	01	0.5	0.5	0.01	02	2.0	1.2	−10.6770
0.5	01	02	01	01	0.5	0.5	0.01	02	2.0	1.2	−10.6946
0.5	01	03	01	01	0.5	0.5	0.01	02	2.0	1.2	−10.7235
0.5	01	05	01	01	0.5	0.5	0.01	02	2.0	1.2	−10.8151
0.5	01	05	02	01	0.5	0.5	0.01	02	2.0	1.2	−10.9559
0.5	01	05	03	01	0.5	0.5	0.01	02	2.0	1.2	−11.0960
0.5	01	05	01	01	0.5	0.5	0.01	02	2.0	1.2	−10.8151
0.5	01	05	01	02	0.5	0.5	0.01	02	2.0	1.2	−10.8064
0.5	01	05	01	03	0.5	0.5	0.01	02	2.0	1.2	−10.7978
0.5	01	05	01	01	2.0	0.5	0.01	02	2.0	1.2	−10.5643
0.5	01	05	01	01	2.5	0.5	0.01	02	2.0	1.2	−10.4735
0.5	01	05	01	01	3.0	0.5	0.01	02	2.0	1.2	−10.3631
0.5	01	05	01	01	0.5	0.3	0.01	02	2.0	1.2	−10.6729
0.5	01	05	01	01	0.5	0.5	0.01	02	2.0	1.2	−10.8151
0.5	01	05	01	01	0.5	0.7	0.01	02	2.0	1.2	−10.9575
0.5	01	05	01	01	0.5	0.5	0.1	02	2.0	1.2	−12.1179
0.5	01	05	01	01	0.5	0.5	0.2	02	2.0	1.2	−14.0721
0.5	01	05	01	01	0.5	0.5	0.3	02	2.0	1.2	−16.1551
0.5	01	05	01	01	0.5	0.5	0.01	10	2.0	1.2	−11.3132
0.5	01	05	01	01	0.5	0.5	0.01	50	2.0	1.2	−15.1132
0.5	01	05	01	01	0.5	0.5	0.01	100	2.0	1.2	−20.1335
0.5	01	05	01	01	0.5	0.5	0.01	02	01	1.2	−10.4104
0.5	01	05	01	01	0.5	0.5	0.01	02	02	1.2	−10.3151
0.5	01	05	01	01	0.5	0.5	0.01	02	03	1.2	−11.2584
0.5	01	05	01	01	0.5	0.5	0.01	02	2.0	0.5	−10.8071
0.5	01	05	01	01	0.5	0.5	0.01	02	2.0	1.0	−10.8127
0.5	01	05	01	01	0.5	0.5	0.01	02	2.0	1.5	−10.8189

7. Conclusions

The axisymmetric flow of Maxwell fluid between two isothermal stretching disks is discussed in presence of source/sink and activation energy features. The mixed convection effects are implemented in the momentum equation. Analytical results are discussed by using the homotopy analysis method. The following observations are furnished:

- The wall shear stress decreases by increasing stretching parameter, Hartmann number, Reynolds number, Deborah number, activation energy parameter and constant temperature parameter. It means that tangential stresses increase by increasing stretching the ratio parameter, Hartmann number and Reynolds number. While the behavior of dimensionless distance and Frank–Kamenetskii number are quite the opposite.
- The pressure distribution is increased with variation of theFrank–Kamenetskii number and stretching ratio parameter.
- When the Deborah number λ and Hartmann number increases, the wall shear stress at the lower disk increases while an opposite trend is found at the upper disk.
- It is observed that the surface heat transfer increases by increasing the stretching parameter and heat source/sink parameter.
- The rate of heat transfer decreases at the lower disk and increases at the upper disk by increasing the Hartmann number, Reynolds number, Archimedes number and activation energy parameter.

Author Contributions: All authors contributed equally. All authors have read and agreed to the published version of the manuscript.

Funding: There is no funding for this work.

Conflicts of Interest: Authors declare no conflict of interest.

Nomenclature

(r, θ, z)	cylindrical coordinate
(u, w)	velocity components
T_1	upper disk temperature
E	activation energy
R_1	gas constant
a_0	reactant concentration
ρ	characteristic density
ν	is the kinematic viscosity
K_T	thermal conductivity of fluid
Q	exothermicity factor
γ	stretching rate constant
M	Hartmann number
α	heat source/sink parameter
Ec	Eckert number
R_T	constant temperature parameter
A_r	Archimedes number
τ_w	shear stress
d	distance
T	is the fluid temperature
T_2	lower disk temperature
B	product species
k_0	is chemical reaction
λ_1	is the relaxation time
p	is the fluid pressure
T_0	isreference temperature
β	denotes the thermal expansion
Q_0	heat generation/absorption coefficient
R	Reynolds number
G_r	Grashoff number
Pr	Prandtl number
K	Frank–Kamenetskii number
ϵ	activation energy parameter
δ	dimensionless distance

References

1. Maleque, K. Effects of exothermic/endothermic chemical reactions with Arrhenius activation energy on MHD free convection and mass transfer flow in presence of thermal radiation. *J. Thermodyn.* **2013**, *2013*, 692516. [CrossRef]
2. Khan, M.I.; Hayat, T.; Khan, M.I.; Alsaedi, A. Activation energy impact in nonlinear radiative stagnation point flow of Cross nanofluid. *Int. Commun. Heat Mass Transf.* **2018**, *91*, 216–224. [CrossRef]
3. Shafique, Z.; Mustafa, M.; Mushtaq, A. Boundary layer flow of Maxwell fluid in rotating frame with binary chemical reaction and activation energy. *Results Phys.* **2016**, *6*, 627–633. [CrossRef]
4. Awad, F.G.; Motsa, S.; Khumalo, M. Heat and Mass Transfer in Unsteady Rotating Fluid Flow with Binary Chemical Reaction and Activation Energy. *PLoS ONE* **2014**, *9*, e107622. [CrossRef] [PubMed]
5. Hsiao, K.L. To promote radiation electrical MHD activation energy thermal extrusion manufacturing system efficiency by using Carreau-Nanofluid with parameters control method. *Energy* **2017**, *130*, 486–499. [CrossRef]

6. Merkin, J.; Mahmood, T. Convective flows on reactive surfaces in porous media. *Transp. Porous Media* **1998**, *47*, 279–293. [CrossRef]
7. Minto, B.J.; Ingham, D.B.; Pop, I. Free convection driven by an exothermic on a vertical surface embedded in porous media. *Int. J. Heat Mass Trans. Transf.* **1998**, *41*, 11–23. [CrossRef]
8. Chou, S.F.; Tsern, I.P. Mixed convection heat transfer of horizontal channel flow over a heated block. *Transp. Phenom. Heat Mass Trans.* **1992**, *2*, 492–503.
9. Nadeem, S.; Khan, M.R.; Khan, A.U. MHD stagnation point flow of viscous nanofluid over a curved surface. *Phys. Scr.* **2019**, *94*, 115207. [CrossRef]
10. Ahmed, Z.; Al-Qahtani, A.; Nadeem, S.; Saleem, S. Computational Study of MHD Nanofluid Flow Possessing Micro-Rotational Inertia over a Curved Surface with Variable Thermophysical Properties. *Processes* **2019**, *7*, 387. [CrossRef]
11. Khan, A.U.; Hussain, S.T.; Nadeem, S. Existence and stability of heat and fluid flow in the presence of nanoparticles along a curved surface by mean of dual nature solution. *Appl. Math. Comput.* **2019**, *353*, 66–81. [CrossRef]
12. Sadiq, M.A.; Khan, A.U.; Saleem, S.; Nadeem, S. Numerical simulation of oscillatory oblique stagnation point flow of a magneto micropolar nanofluid. *RSC Adv.* **2019**, *9*, 4751–4764. [CrossRef]
13. Harris, J. *Rheology and Non-Newtonian Flow*; Longman Inc.: New York, NY, USA, 1977.
14. Hayat, T.; Abbas, Z.; Sajid, M. Series Solution for the Upper-Convected Maxwell Fluid over a Porous Stretching Plate. *Phys. Lett. A* **2006**, *358*, 396–403. [CrossRef]
15. Aliakbar, V.; Pahlavan, A.A.; Sadeghy, K. The Influence of Thermal Radiation on MHD Flow of Maxwellian Fluid above Stretching Sheet. *Commun. Nonlinear Sci. Numer. Simul.* **2009**, *14*, 779–794. [CrossRef]
16. Hayat, T.; Abbas, Z.; Sajid, M. MHD Stagnation-point Flow of an Upper-Convected Maxwell Fluid over a Stretching Surface. *ChaosSolitons Fractals* **2009**, *39*, 840–848. [CrossRef]
17. Prasad, K.V.; Sujatha, A.; Vajravelu, K.; Pop, I. MHD Flow and Heat Transfer of a UCM Fluid over a Stretching Surface with Variable Thermophysical Properties. *Meccanica* **2012**, *47*, 1425–1439. [CrossRef]
18. Khan, S.U.; Ali, N.; Sajid, M.; Hayat, T. Heat transfer characteristics in oscillatory hydromagnetic channel flow of Maxwell fluid using Cattaneo-Christov model. *Proc. Natl. Acad. Sci. India Sect. A Phys. Sci.* **2019**, *89*, 377–385. [CrossRef]
19. Ahmed, J.; Khan, M.; Ahmad, L. MHD swirling flow and heat transfer in Maxwell fluid driven by two coaxially rotating disks with variable thermal conductivity. *Chin. J. Phys.* **2019**, *60*, 22–34. [CrossRef]
20. Merkin, J.H.; Chaudhary, M.A. Free convection boundary layer driven by an exothermic surface reaction. *Q. J. Mech. Appl. Math.* **1994**, *47*, 405–428. [CrossRef]
21. Gorder, R.A.V.; Sweet, E.; Vajravelu, K. Analytical solutions of a coupled nonlinear system arising in a flow between stretching disks. *Appl. Math. Comput.* **2010**, *216*, 1513–1523. [CrossRef]
22. Khan, N.; Mahmood, T.; Sajid, M.; Hashmi, M.S. Heat and mass transfer on MHD mixed convection axisymmetric chemically reactive flow of Maxwell fluid driven by exothermal and isothermal stretching disks. *Int. J. Heat Mass Transf.* **2016**, *92*, 1090–1105. [CrossRef]
23. Liao, S.J. On the analytic solution of magnetohydrodynamic flow of non-Newtonian fluids over a stretching sheet. *J. Fluid Mech.* **2003**, *488*, 189–212. [CrossRef]
24. Turkyilmazoglu, M. Solution of the Thomas–Fermi equation with a convergent approach. *Commun. Nonlinear Sci. Numer. Simul.* **2012**, *17*, 4097–4103. [CrossRef]
25. Turkyilmazoglu, M. ParametrizedAdomian decomposition method with optimum convergence. *ACM Trans. Modeling Comput. Simul.* **2017**, *24*, 21.
26. Khan, S.U.; Ali, N.; Hayat, T. Analytical and Numerical Study of Diffusion of Chemically Reactive Species in Eyring-Powell Fluid over an Oscillatory Stretching Surface. *Bulg. Chem. Commun.* **2017**, *49*, 320–330.
27. Ullah, I.; Waqas, M.; Hayat, T. Thermally radiated squeezed flow of magneto-nanofluid between two parallel disks with chemical reaction. *J. Therm. Anal. Calorim.* **2019**, *135*, 1021. [CrossRef]
28. Khan, S.U.; Shehzad, S.A.; Ali, N. Darcy-Forchheimer MHD Couple Stress liquid flow by oscillatory stretched sheet with thermophoresis and heat generation/absorption. *J. Porous Media* **2018**, *21*, 1197–1213. [CrossRef]
29. Ellahi, R.; Hassan, M.; Zeeshan, A. Aggregation effects on water base Al_2O_3—Nano fluid over permeable wedge in mixed convection. *Asia-Pac. J. Chem. Eng.* **2016**, *11*, 179–186. [CrossRef]
30. Hassan, M.; Ellahi, R.; Zeeshan, A.; Bhatti, M.M. Analysis of natural convective flow of non-Newtonian fluid under the effects of nanoparticles of different materials. *J. Process Mech. Eng.* **2019**, *233*, 643–652. [CrossRef]

31. Ambreen, A.K.; Bukhari, S.R.; Marin, M.; Ellahi, R. Effects of chemical reaction on third grade magnetohydrodynamics fluid flow under the influence of heat and mass transfer with variable reactive index. *Heat Transf. Res.* **2019**, *50*, 1061–1080.
32. Sajjad, R.; Hayat, T.; Ellahi, R.; Muhammad, T.; Alsaedi, A. Darcy–Forchheimer flow of nanofluid due to a curved stretching surface. *Int. J. Numer. Methods Heat Fluid Flow* **2019**, *29*, 2–20.
33. Hassan, M.; Ellahi, R.; Bhatti, M.M.; Zeeshan, A. A comparitive study of magnetic and non-magnetic particles in nanofluidpropigating over over a wedge. *Can. J. Phys.* **2019**, *97*, 277–285. [CrossRef]
34. Hussain, F.; Ishtiaq, F.; Hussain, A. Peristaltic transport of Jeffrey fluid in a rectangular duct through a porous medium under the effect of partial slip: An application to upgrade industrial sieves/filters. *Pramana J. Phys.* **2019**, *93*, 34.
35. Ellahi, R.; Hassan, M.; Zeeshan, A. A study of heat transfer in power law nanofluid. *Therm. Sci. J.* **2016**, *20*, 2015–2026. [CrossRef]
36. Zeeshan, A.; Hassan, M.; Ellahi, R.; Nawaz, M. Shape effect of nanosize particles in unsteady mixed convection flow of nanofluid over disk with entropy generation. *Proc. Inst. Mech. Eng. Part E J. Process Mech. Eng.* **2017**, *231*, 871–879. [CrossRef]

Article

Keller-Box Analysis of Buongiorno Model with Brownian and Thermophoretic Diffusion for Casson Nanofluid over an Inclined Surface

Khuram Rafique [1], Muhammad Imran Anwar [1,2,3], Masnita Misiran [1], Ilyas Khan [4,*], Sayer O. Alharbi [5], Phatiphat Thounthong [6] and Kottakkaran Sooppy Nisar [7]

1 School of Quantitative Sciences, Universiti Utara Malaysia, Sintok 06010, Kedah, Malaysia; Khurram.rafique1005@gmail.com (K.R.); masnita@uum.edu.my (M.M.)
2 Department of Mathematics, Faculty of Science, University of Sargodha, Sargodha 40411, Pakistan
3 Higher Education Department (HED), Punjab 54000, Pakistan; imrananwar@uos.edu.pk
4 Faculty of Mathematics and Statistics, Ton Duc Thang University, Ho Chi Minh City 72915, Vietnam
5 Department of Mathematics, College of Science Al-Zulfi, Majmaah University, Al-Majmaah 11952, Saudi Arabia; so.alharbi@mu.edu.sa
6 Renewable Energy Research Centre, Department of Teacher Training in Electrical Engineering Faculty of Technical Education King Mongkut's University of Technology North Bangkok, 1518, Pracharat 1 Rd., Bangsue, Bangkok 10800, Thailand; phatiphat.t@fte.kmutnb.ac.th
7 Department of Mathematics, College of Arts and Science, Prince Sattam bin Abdulaziz University, Wadi Al-Dawaser 11991, Saudi Arabia; n.sooppy@psau.edu.sa
* Correspondence: ilyaskhan@tdtu.edu.vn

Received: 10 August 2019; Accepted: 17 September 2019; Published: 5 November 2019

Abstract: The key objective of the study under concern is to probe the impacts of Brownian motion and thermophoresis diffusion on Casson nanofluid boundary layer flow over a nonlinear inclined stretching sheet, with the effect of convective boundaries and thermal radiations. Nonlinear ordinary differential equations are obtained from governing nonlinear partial differential equations by using compatible similarity transformations. The quantities associated with engineering aspects, such as skin friction, Sherwood number, and heat exchange along with various impacts of material factors on the momentum, temperature, and concentration, are elucidated and clarified with diagrams. The numerical solution of the present study is obtained via the Keller-box technique and in limiting sense are reduced to the published results for accuracy purpose.

Keywords: Keller-box technique; Casson nanofluid; MHD; Power law fluid; Convective boundaries; Radiation effect; Inclined surface

1. Introduction

Brownian motion and thermophoresis diffusions are the key notions of abnormal improvement in thermal conductivity by using binary fluids (base fluid along with nanoparticles). The influence of Brownian motion and thermophoresis is focused in the Buongiorno model. This model supports engineers and scholars through its utilization in the field of science and technology. It is also pointed out that nanoparticles occupying Brownian motion and thermophoresis effects cause improvement of thermal conductivity. The Brownian motion principle along thermophoresis particle installation supports, in manufacturing, germanium dioxide optical fibers and, in communication engineering, silicon. The impacts of Brownian motion and thermophoresis diffusion on Casson nanofluid flow on a stretching sheet were discussed by Anwar et al. [1]. Afify [2] scrutinized the Brownian movement and thermophosis impact on Casson nanofluid flow with convective boundaries. The impacts of radiations on Casson nanofluid flow with Brownian motion and thermophoresis influence were studied by

Souayeh et al. [3]. Rashidi et al. [4] discussed heat exchange and particle motion by considering the discrete phase model (DPM). Bhatti et al. [5] examined electro-magnetohydrodynmaic (MHD) flow with heat exchange by incorporating the thermal radiations effect. Ellahi et al. [6] investigated a shiny thin film with metallic tactile covering nanoparticles through a rotating disk. Numerous scholars [7–12] considered the Buongiorno model to investigate flow characteristics.

The most significant concerns of a creator and craftsman in the construction of different items, in the pursuit of excellence, is the lessening of expenses and time. The role of heat and fluids in industries is undeniable. Discovering approaches towards the advancement of procedures and the quantity of energy exchange has consistently been a concern for researchers and specialists from the earlier times to the current era. The discovery of nanoparticles and the advancement in nanotechnology is viewed as a tremendous change in innovation and science. Choi [13] was the initiator of the nanofluid concept. A mixture of a base fluid (water, ethylene glycol and so on) with nano-scale particles called nanoparticles is termed as a nanofluid. Nanofluids have a higher thermal conductivity as compared to base fluids, due to which the energy exchange procedure is enhanced. The radiation effects on Casson nanofluid flow on a nonlinear slanted sheet were investigated by Ghadikolaei et al. [14]. The effect of a magnetic field on the flow of a nanofluid over an inclined sheet was studied by Suriyakumar and Devi [15]. Khan et al. [16] examined the flow of a Jeffery nanofluid over a slanted sheet. Thumma et al. [17] discussed the flow of a nanofluid over a nonlinear inclined stretching sheet. Parkash et al. [18] discussed the nanofluid flow through a channel analytically. Zeeshan et al. [19] examined the flow of titanium dioxide-water base nanofluid because of entropy generation. Shehzad et al. [20] calculated the silver-water base nanofluid flow in a porous medium because of entropy generation. Hussain et al. [21] studied multiphase flow synthesis with nano-size hafnium particles with the effect of electro-hydrodynamic effect. Ellahi et al. [22] discussed the thermally charged MHD bi-phase flow coating with non-Newtonian nanofluid along slipper walls. Recently, many scholars discussed nanofluid flow by incorporating different impacts [23–26].

Non-Newtonian fluids have gained considerable attention from scientists and engineers because of their key role in the field of industry and engineering. The study under concern has direct noteworthy use in association with non-Newtonian fluids, such as Casson fluids (honey, human blood etc.), power law fluids and nanofluids etc. Reddy [27] investigated the movement of Casson liquid over a slanted sheet. Hakeem et al. [28] investigated the inclined Lorentz force on Casson fluid flow on an extended sheet. Rawi et al. [29] studied the unsteady flow of Casson fluid through a slanted sheet. Casson fluid flow on an inclined sheet with multiple impacts was discussed by Jain and Parmar [30]. Ellahi et al. [31] discussed the two-phase Couette flow of couple stress by incorporating the magnetic field impacts. Ellahi et al. [32] investigated the blood flow of couple stress fluid with chemical reaction effects. For further detailed literature related to non-Newtonian fluid flow on inclined sheets, see [33–40].

Heat exchange due to thermal radiations has become an active area of research due to its vast range of applications in the field of nuclear power plants, missiles, satellites and in nanotechnology. Moreover, it is significant that thermal radiation is not suitable for the engineering of thermal tools with large variations in temperature [41]. The thermal radiation impact on flow and heat exchange is a key factor to design advanced energy conversion systems [42]. Recently, Ghadikolaei et al. [43] investigated the flow of Casson nanofluid on a porous inclined sheet numerically. Saidulu [44] discussed the radiation impacts on the flow of a nanofluid over an exponential inclined surface.

To the best of the authors' knowledge, no study on Casson nanofluid flow over an inclined nonlinear stretching sheet along with radiation effects and convective boundaries has been reported yet. Besides, the fact that a lot of work has already been done on non-Newtonian fluids with different geometries, but due to the growing applications of non-Newtonian fluids in the field of industry, the authors choose this study on an inclined sheet. The non-Newtonian fluid flow on an inclined sheet plays a vital role in MHD generators, gas turbines, and extrusion of plastic sheets. The numerical solution of the current problem is obtained using the Keller-box method.

2. Problem Formulation

Suppose two-dimensional incompressible Casson nanofluid flow over a nonlinear inclined stretching sheet slanted at γ, where $u_w(x) = ax^m$ is the extending speed and $u_\infty(x) = 0$ is free stream speed, in which x is the coordinate stated towards the extending sheet and 'a' is considered as constant. The transverse magnetic field 'B_0' is taken as normal to the track of flow. The Brownian motion and thermophoresis effects are considered. The temperature T and nanoparticle fraction C take the values T_w and C_w at the wall. The thermal radiation impact is incorporated with a convective heating procedure considered by the temperature T_f and heat exchange factor h_f, which is proportional to x^{-1}. Meanwhile, the encompassing structures for nanofluid temperature and mass divisions T_∞ and C_∞ are achieved as y keeps an eye on infinity, as displayed in Figure 1.

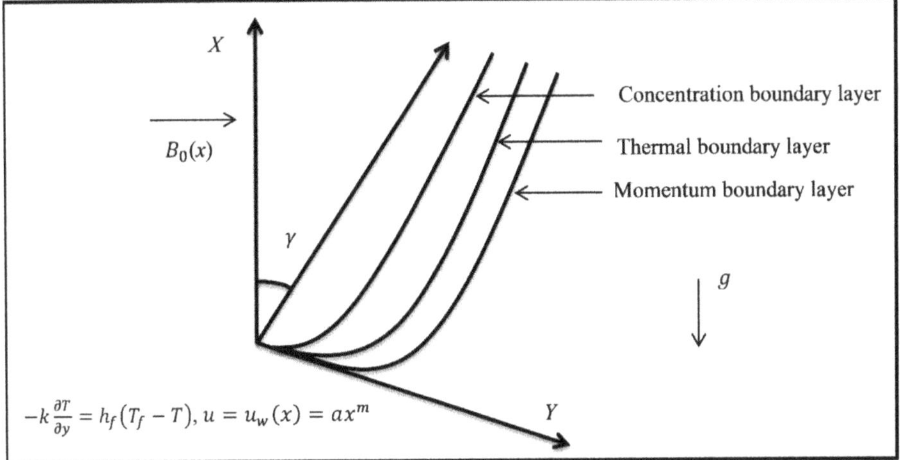

Figure 1. Physical geometry with coordinate system.

The flow equations for this study are given by

$$\frac{\partial u}{\partial x} + \frac{\partial v}{\partial y} = 0, \tag{1}$$

$$u\frac{\partial u}{\partial x} + v\frac{\partial u}{\partial y} = \nu\left(1 + \frac{1}{\beta}\right)\frac{\partial^2 u}{\partial y^2} + g[\beta_t(T - T_\infty) + \beta_c(C - C_\infty)]cos\gamma - \frac{\sigma {B_0}^2(x)}{\rho}u, \tag{2}$$

$$u\frac{\partial T}{\partial x} + v\frac{\partial T}{\partial y} = \alpha\frac{\partial^2 T}{\partial y^2} - \frac{1}{(\rho c)_f}\frac{\partial q_r}{\partial y} + \tau\left[D_B\frac{\partial C}{\partial y}\frac{\partial T}{\partial y} + \frac{D_T}{T_\infty}\left(\frac{\partial T}{\partial y}\right)^2\right], \tag{3}$$

$$u\frac{\partial C}{\partial x} + v\frac{\partial C}{\partial y} = D_B\frac{\partial^2 C}{\partial y^2} + \frac{D_T}{T_\infty}\frac{\partial^2 T}{\partial y^2}. \tag{4}$$

Here, the Rosseland estimation (for radiation flux) is characterized as

$$q_r = -\frac{4\sigma^*}{3k^*}\frac{\partial T^4}{\partial y} \tag{5}$$

where the Stephen-Boltzmann coefficient is given by σ^* and the mean absorption constant is represented by k^*. Meanwhile, the temperature changes between the local temperature T and free steam T_∞ are very small, by ignoring higher order terms in the expansion of T^4 in Taylor succession about T_∞ for:

$$T^4 \cong 4T_\infty^3 T - 3T_\infty^4. \tag{6}$$

By using Equations (5) and (6), the Equation (3) is converted into

$$u\frac{\partial T}{\partial x} + v\frac{\partial T}{\partial y} = \left(\alpha + \frac{16\sigma^* T_\infty^3}{3k^*(\delta c)_f}\right)\frac{\partial^2 T}{\partial y^2} + \tau\left[D_B \frac{\partial C}{\partial y}\frac{\partial T}{\partial y} + \frac{D_T}{T_\infty}\left(\frac{\partial T}{\partial y}\right)^2\right], \tag{7}$$

where in the directions x and y, the velocity constituents are u and v, individually, g is the gravitational acceleration, the strength of the magnetic field is defined by B_0, σ is the electrical conductivity, viscosity is given by μ, the density of conventional fluid is given by ρ_f, the density of the nanoparticle is given by ρ_p, β is the Casson parameter, the thermal expansion factor is denoted by β_t, the concentration expansion constant is given by β_c, D_B denotes the Brownian dissemination factor and D_T represents the thermophoresis dispersion factor. The thermal conductivity is given by k, the heat capacity of the nanoparticles symbolically is given as $(\rho c)_p$, the heat capacity of the conventional liquid is given by $(\rho c)_f, \alpha = \frac{k}{(\rho c)_f}$ denotes the thermal diffusivity parameter, and the symbolic representation of the relation among the current heat capacity of the nanoparticle and the liquid is $\tau = \frac{(\rho c)_p}{(\rho c)_f}$.

In this problem, boundary conditions are considered as

$$\begin{gathered} u = u_w(x) = ax^m,\ v = 0,\ -k\tfrac{\partial T}{\partial y} = h_f\left(T_f - T\right),\ C = C_w\ at\ y = 0, \\ u \to u_\infty(x) = 0,\ v \to 0,\ T \to T_\infty,\ C \to C_\infty\ at\ y \to \infty \end{gathered} \tag{8}$$

For the conversion of the Equations (2), (4) and (7) into ordinary differential equations, we use $\psi = \psi(x, y)$, called the stream function, characterized as

$$u = \frac{\partial \psi}{\partial y},\quad v = -\frac{\partial \psi}{\partial x}. \tag{9}$$

The similarity transformations are considered as

$$\psi = \sqrt{\frac{2v a x^{m+1}}{m+1}} f(\eta),\ \theta(\eta) = \frac{T - T_\infty}{T_w - T_\infty}, \phi(\eta) = \frac{C - C_\infty}{C_w - C_\infty},\ \eta = y\sqrt{\frac{(m+1)ax^{m-1}}{2v}} \tag{10}$$

By using Equations (9) and (10), Equation (1) is fulfilled indistinguishably. Besides, Equations (2), (4) and (7) are transformed to the following

$$\left(1 + \frac{1}{\beta}\right)f''' + ff'' - \left(\frac{2m}{m+1}\right)f'^2 + \frac{2}{m+1}(\lambda\theta + \delta\phi)cos\gamma - \left(\frac{2M}{m+1}\right)f' = 0 \tag{11}$$

$$Pr_N \theta'' + f\theta' + Nb\phi'\theta' + Nt\theta'^2 = 0 \tag{12}$$

$$\phi'' + Lef\phi' + Nt_b\theta'' = 0 \tag{13}$$

where

$$\begin{gathered} \lambda = \tfrac{Gr_x}{Re_x{}^2},\ \delta = \tfrac{Gc_x}{Re_x{}^2},\ M = \tfrac{\sigma B_0{}^2(x)}{a\rho},\ Le = \tfrac{v}{D_B} Pr = \tfrac{v}{\alpha},\ N_b = \tfrac{\tau D_B(C_w - C_\infty)}{v}, \\ N_t = \tfrac{\tau D_t(T_w - T_\infty)}{v T_\infty},\ Nt_b = \tfrac{N_t}{N_b},\ Gr_x = \tfrac{g\beta_t(T_w - T_\infty)x^3}{v^2},\ Re_x = \tfrac{u_w x}{v},\ Gc_x = \tfrac{g\beta_c(C_w - C_\infty)x^3}{v^2}, \\ Pr_N = \tfrac{1}{Pr}\left(1 + \tfrac{4}{3}N\right),\ N = \tfrac{4\sigma^* T_\infty^3}{\alpha k^*}. \end{gathered} \tag{14}$$

Here, primes signify the differentiation concerning η, λ is the buoyancy parameter, δ is the solutal buoyancy parameter, the magnetic constraint is given by M, v denotes the kinematic viscosity of the liquid, the Prandtl number is given as Pr, the Lewis number is given by Le, N_b denotes the Brownian motion parameter, N_t indicates the thermophoresis factor and N denotes the radiation factor.

The resultant boundary settings are

$$\begin{gathered} f(\eta)=0,\ f'(\eta)=1,\ \theta'(0)=-\gamma_1(1-\theta(0)),\ \phi(\eta)=1 \ at\ \eta=0, \\ f'(\eta)\to 0,\ \theta(\eta)\to 0,\ \phi(\eta)\to 0 \ as\ \eta\to\infty. \end{gathered} \tag{15}$$

Here, $\gamma_1=\frac{n}{k\sqrt{Re_x}}$ is the convective parameter termed as Biot number.

The skin friction, Sherwood number and Nusselt number for the current study are regarded as

$$Nu_x=\frac{xq_w}{k(T_w-T_\infty)},\ Sh_x=\frac{xq_m}{D_B(C_w-C_\infty)},\ C_f=\frac{t_w}{u_w{}^2\rho_f}, \tag{16}$$

where

$$q_w=-\left[k+\frac{4\sigma^*T_\infty^3}{3k^*},\right]\frac{\partial T}{\partial y},\ q_m=-D_B\frac{\partial C}{\partial y},\ \tau_w=\mu\left(1+\frac{1}{\beta}\right)\frac{\partial u}{\partial y},\ at\ y=0.$$

The related terms of dimensionless reduced Nusselt number $-\theta'(0)$, reduced Sherwood number $-\phi'(0)$ and skin friction coefficient $C_{fx}=\left(1+\frac{1}{\beta}\right)f''(0)$ are defined as

$$-\theta'(0)=\frac{Nu_x}{\left(1+\frac{4}{3}N\right)\sqrt{\frac{m+1}{2}Re_x}},-\phi'(0)=\frac{Sh_x}{\sqrt{\frac{m+1}{2}Re_x}},C_{fx}=C_f\sqrt{\frac{2}{m+1}Re_x}, \tag{17}$$

where $Re_x=\frac{u_w x}{v}$, is the local Reynolds number.

The converted nonlinear differential Equations (11)–(13) with the boundary settings (15) are elucidated by a Keller-box scheme consisting of the steps as finite-differences scheme, Newton's technique and block elimination process, clearly explained by Anwar et al. [7]. The Keller-box technique has been widely applied because it is the most flexible as compared to other approaches. It is informal to practice, much quicker, friendly to program and effective.

3. Results and Discussion

In this part of the study, the numerical outcomes of the converted nonlinear ordinary differential Equations (11)–(13) with boundary settings (15) are elucidated by the Keller-box method. For the numerical results of physical parameters of our concern, namely, Brownian motion denoted by Nb, thermophoresis given by Nt, magnetic factor M, buoyancy factor λ, solutal buoyancy constraint δ, inclination factor γ, Prandtl number Pr, Lewis number Le, radiation factor N, Casson fluid parameter β, Biot number γ_1 and parameter m, several figures and tables are prepared. In Table 1, in the deficiency of λ, δ, M, and N, and taking factor $m=1$, with $\gamma = 90°$ and $\beta\to\infty$, the outcomes of $-\theta'(0)$, $-\phi'(0)$ (reduced Nusselt number, reduced Sherwood number) are equated with the results of Khan and Pop [45]. The magnitudes are established as brilliant settlement. The effects on $-\theta'(0)$, $-\phi'(0)$ and $C_{fx}(0)$ against several values of involved physical parameters Nb, β, Nt, M, N, λ, δ, γ, Pr, Le, γ_1 and m are presented in Table 2. It is noted that $-\theta'(0)$ drops when increasing the values of Nb, Pr, β, N, Le, m, and γ_1, and it increased by enhancing the numerical values of γ, λ, δ, M and Nt. Moreover, it is perceived that $-\phi'(0)$ is enhanced with larger values of Nb, Pr, N, Le, λ, and δ, and drops for bigger values of m, M, β, γ, γ_1 and Nt. Physically, by enhancing the Brownian motion impact, the thermal boundary layer thickness increases, and it effects a large amount of the fluid. Moreover, the Sherwood number increases and the Nusselt number decreases as we boost the thermophoresis effect; this is due to the fact that the thermal boundary layer turns thicker due to deeper diffusion penetration into the fluid. On the other hand, $C_{fx}(0)$ rises with the growing values of Nb, Pr, Le, β, M, N,γ, and m, and drops with the higher values of Nt, λ, δ, γ_1 and Pr.

Table 1. Contrast of the reduced Nusselt number $-\theta'(0)$ and the reduced Sherwood number $-\phi'(0)$ against $\gamma = 90°$, $\gamma_1 \to \infty$, $\beta \to \infty$, $M, N, \lambda, \delta = 0$, with $m = 1$, and $Pr = Le = 10$.

Nb	Nt	Khan and Pop [45]		Present Results	
		$-\theta'(0)$	$-\phi'(0)$	$-\theta'(0)$	$-\phi'(0)$
0.1	0.1	0.9524	2.1294	0.9524	2.1294
0.2	0.2	0.3654	2.5152	0.3654	2.5152
0.3	0.3	0.1355	2.6088	0.1355	2.6088
0.4	0.4	0.0495	2.6038	0.0495	2.6038
0.5	0.5	0.0179	2.5731	0.0179	2.5731

Table 2. Values of the reduced Nusselt number $-\theta'(0)$, the reduced Sherwood number $-\phi'(0)$ and the skin friction coefficient $C_{fx}(0)$.

Nb	Nt	Pr	Le	M	N	β	λ	δ	γ_1	m	γ	$-\theta'(0)$	$-\phi'(0)$	$C_{fx}(0)$
0.1	0.1	6.5	5.0	0.1	1.0	1.0	0.1	0.9	0.1	0.5	45°	0.0936	1.6159	0.5417
0.5	0.1	6.5	5.0	0.1	1.0	1.0	0.1	0.9	0.1	0.5	45°	0.0447	1.6541	0.5449
0.1	**0.13**	6.5	5.0	0.1	1.0	1.0	0.1	0.9	0.1	0.5	45°	0.0961	1.6038	0.5406
0.1	0.1	**10.0**	5.0	0.1	1.0	1.0	0.1	0.9	0.1	0.5	45°	0.0563	1.7133	0.6235
0.1	0.1	6.5	**10.0**	0.1	1.0	1.0	0.1	0.9	0.1	0.5	45°	0.0919	2.3622	0.5785
0.1	0.1	6.5	5.0	**0.3**	1.0	1.0	0.1	0.9	0.1	0.5	45°	0.0957	1.5958	0.6332
0.1	0.1	6.5	5.0	0.1	**5.0**	1.0	0.1	0.9	0.1	0.5	45°	0.0527	1.6374	0.5421
0.1	0.1	6.5	5.0	0.1	1.0	**5.0**	0.1	0.9	0.1	0.5	45°	0.0916	1.5833	0.6563
0.1	0.1	6.5	5.0	0.1	1.0	1.0	**1.0**	0.9	0.1	0.5	45°	0.0952	1.6184	0.5250
0.1	0.1	6.5	5.0	0.1	1.0	1.0	0.1	**2.0**	0.1	0.5	45°	0.0949	1.6376	0.3809
0.1	0.1	6.5	5.0	0.1	1.0	1.0	0.1	0.9	**0.2**	0.5	45°	0.1779	1.5828	0.5363
0.1	0.1	6.5	5.0	0.1	1.0	1.0	0.1	0.9	0.1	**1.5**	45°	0.1253	1.5763	0.6968
0.1	0.1	6.5	5.0	0.1	1.0	1.0	0.1	0.9	0.1	0.5	**60°**	0.0968	1.6098	0.5818

Figure 2 demonstrates the velocity profile against the magnetic effect. It is observed that the magnetic parameter produces Lorentz force, due to which the velocity of the fluid retards and the velocity profile drops for higher values of the magnetic parameter. Moreover, Figure 3 shows that the temperature contour is enhanced by upgrading the magnetic parameter, and the reason behind this is that the Lorentz force boosts the temperature. Consequently, the thickness of the boundary layer upturns with the increasing of the magnetic parameter. Besides, a different effect of magnetic field on concentration is noticed in Figure 4.

The influence of the nonlinear parameter on the velocity profile is shown in Figure 5. It is noted that the velocity field is not much pronounced in the case of a linear or nonlinear stretching sheet as compared to a uniformly moving surface. Similar behavior is shown in Figure 6 for the temperature profile. Moreover, an opposite effect is shown in Figure 7 for the concentration profile. Figure 8 represents the Casson effect suppressed the velocity of the fluid. It is meaningful because β reduces the yield stress in the Casson fluid. Physically, an enhancement in the Casson parameter tends to reduce the yield stress, which implies that the plastic dynamic viscosity of the liquid is enhanced and the momentum boundary layer becomes thicker [46].

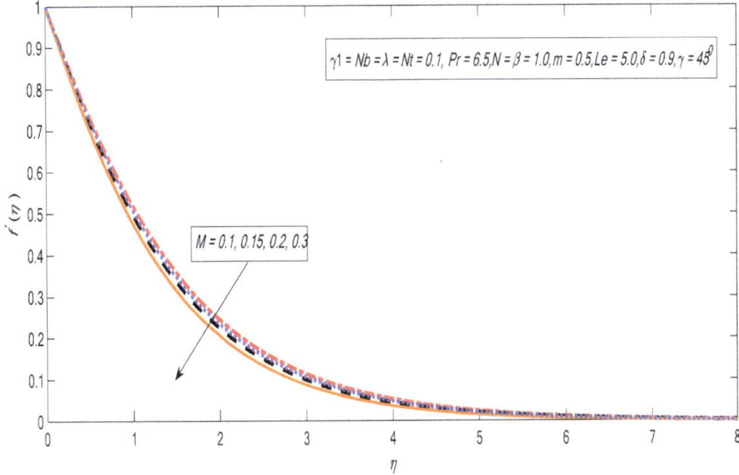

Figure 2. Variations in velocity profile for several values of M.

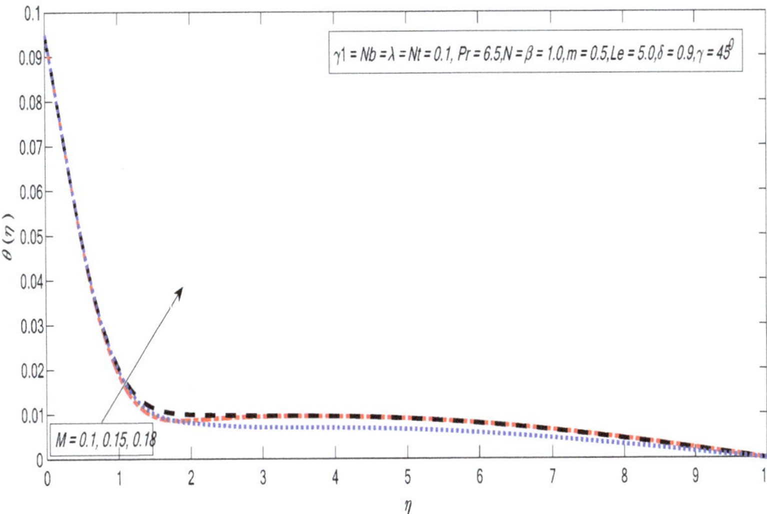

Figure 3. Variations in temperature profile for several values of M.

Figure 9 indicates that the buoyancy force parameter λ has a directly proportional relation with the velocity profile. Physically, an increment in the buoyancy force causes a reduction of the viscous force, due to which the fluid particles move faster. In summary, the enhancement in buoyancy force tends to enhance the velocity profile. Figure 10 reveals the effect of solutal buoyancy impact on the velocity profile. The concentration difference, length and viscosity of the fluid affected the solutal buoyancy parameter. Therefore, as we enhance the solutal buoyancy parameter, the viscosity declines and the concentration increases, due to which the velocity of the fluid increases [47]. Figure 11 reflects the impact of inclination factor γ on the velocity profile. It is observed in Figure 11 that the velocity contour runs down by improving the values of γ. Moreover, the conditions specify that the maximum gravitational force acts on the flow in the case of $\gamma = 0$, because in this state the sheet will be vertical. On the other hand, for $\gamma = 90^{\circ}$, the sheet will be horizontal, which causes a drop in velocity profile as the power of the bouncy forces drop. Figure 12 represents the impact of radiations on the temperature profile. It reveals that the temperature profile increases with large values of thermal radiation parameter; the reason behind this is that the heat exchange is enhanced and the boundary layer thickness declines [48]. Figure 13 indicates that the temperature profile is enhanced near the boundary layer by improving the values of the Biot number. A similar behavior in the case of concentration outline against higher values of the Biot number is seen in Figure 14.

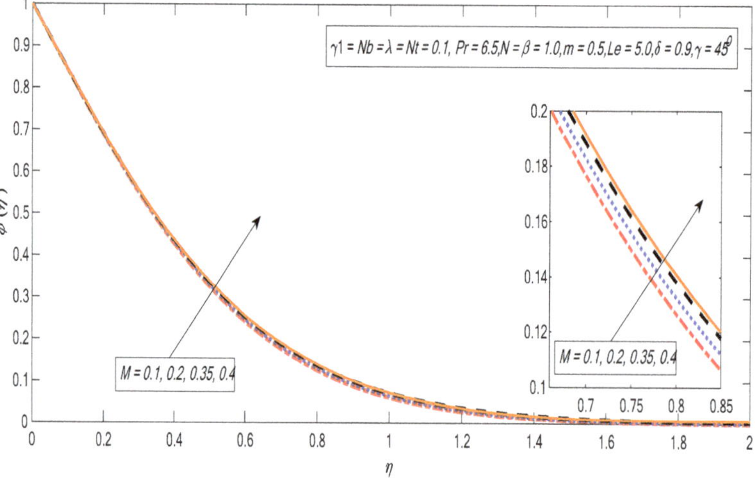

Figure 4. Variations in concentration profile for several values of M.

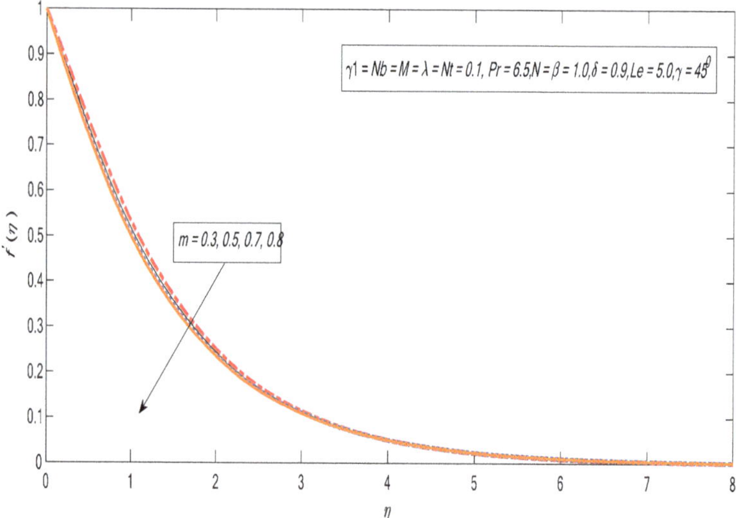

Figure 5. Variations in velocity profile for several values of m.

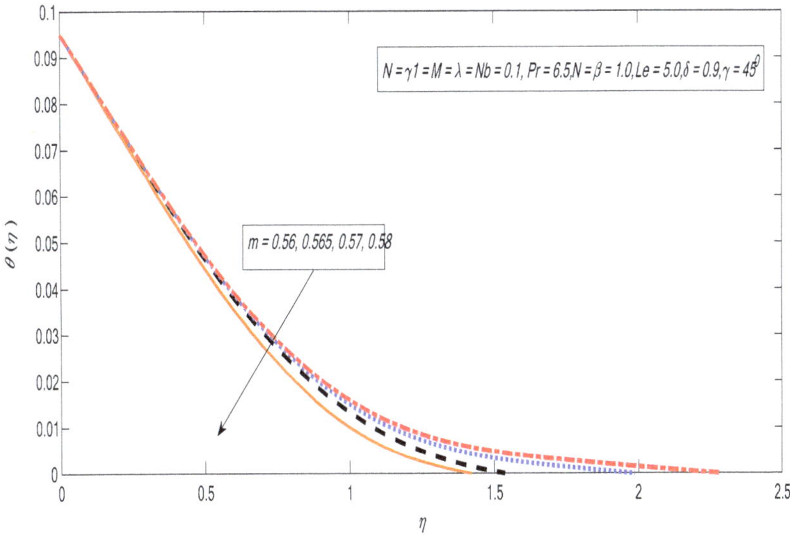

Figure 6. Variations in temperature profile for several values of m.

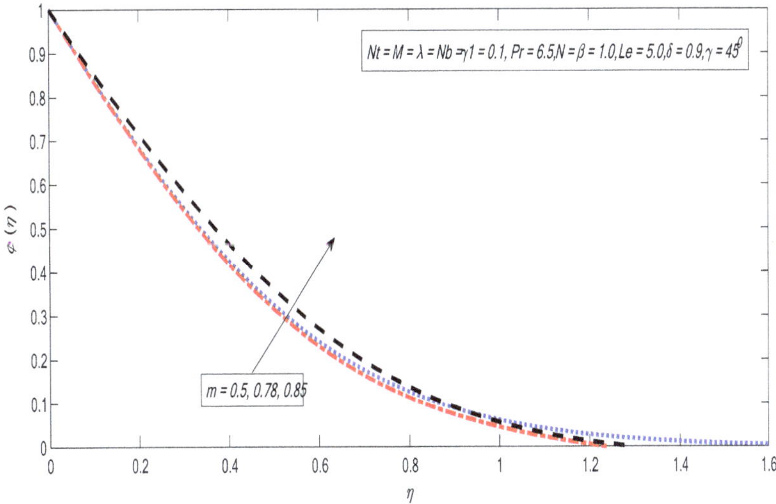

Figure 7. Variations in concentration profile for several values of m.

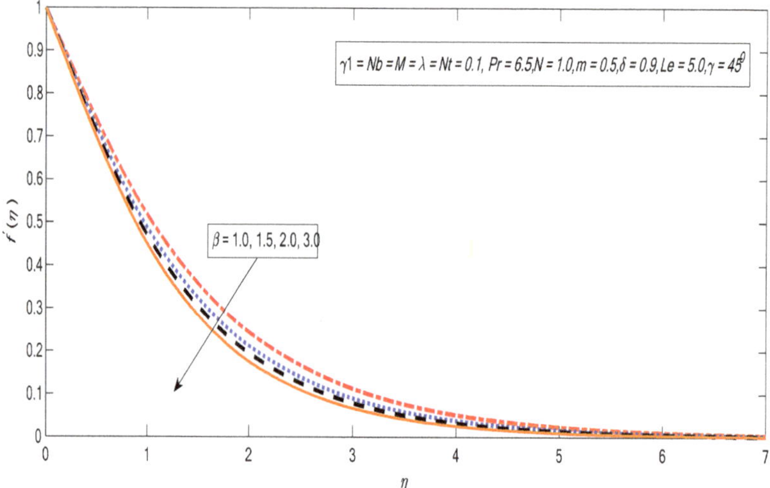

Figure 8. Variations in velocity profile for several values of β.

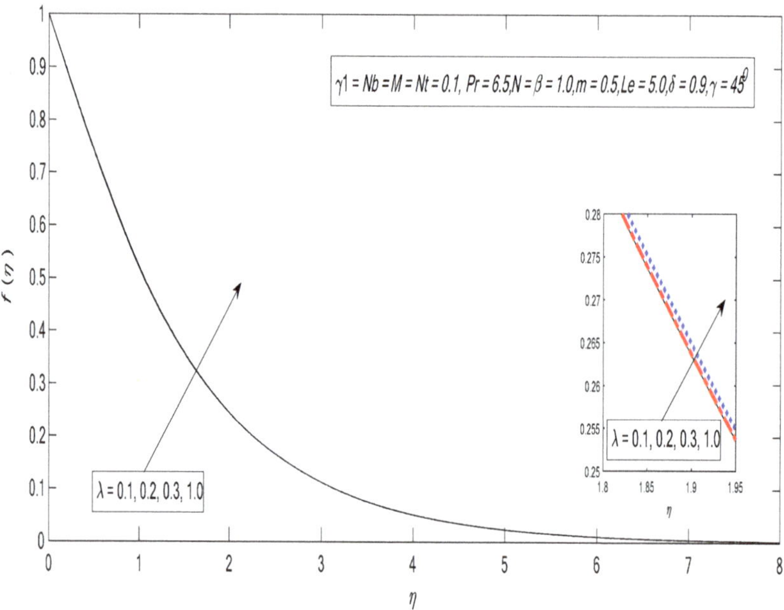

Figure 9. Variations in velocity profile for several values of λ.

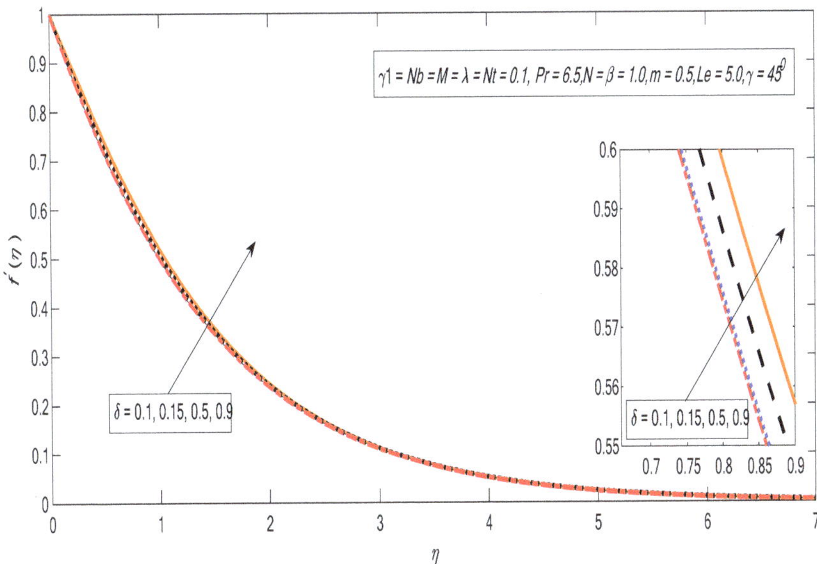

Figure 10. Variations in velocity profile for several values of δ.

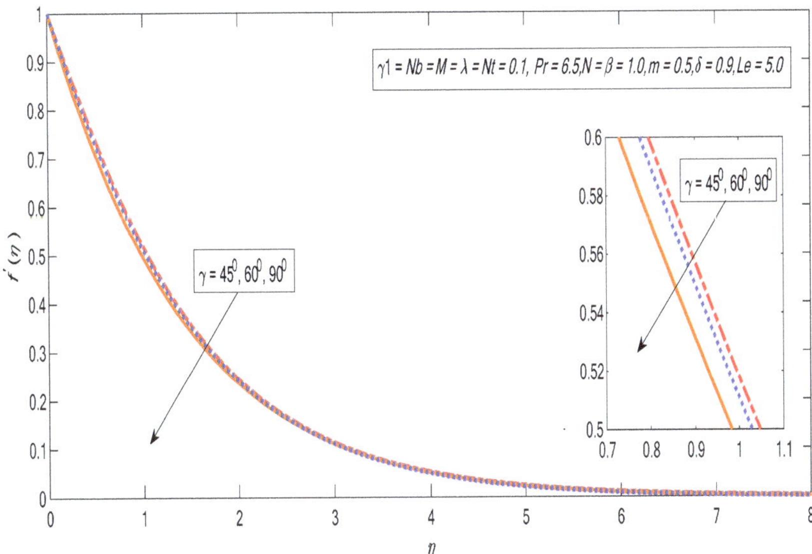

Figure 11. Variations in velocity profile for several values of γ.

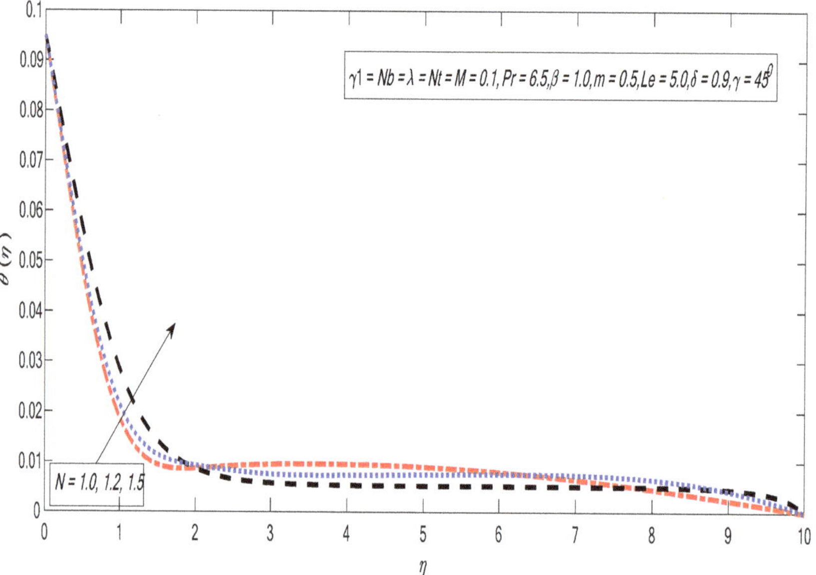

Figure 12. Variations in temperature profile for several values of N.

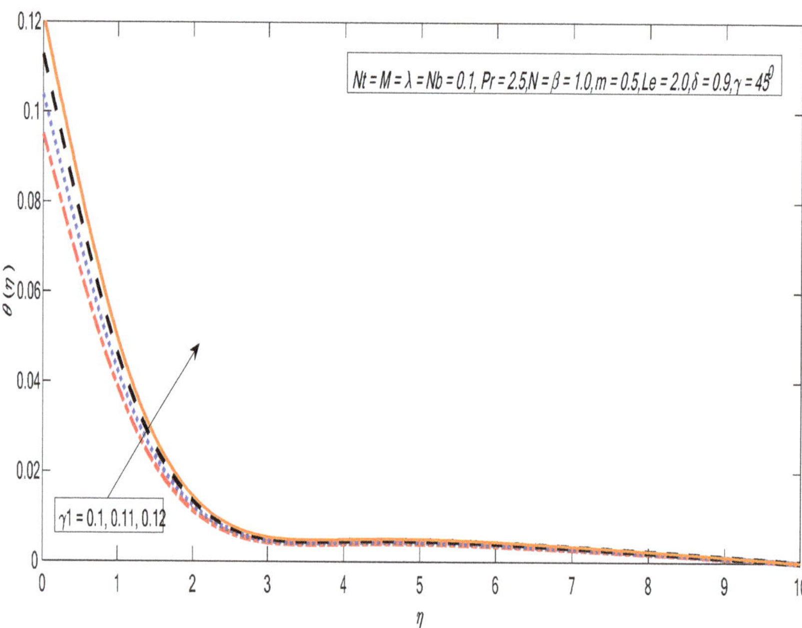

Figure 13. Variations in temperature profile for several values of γ_1.

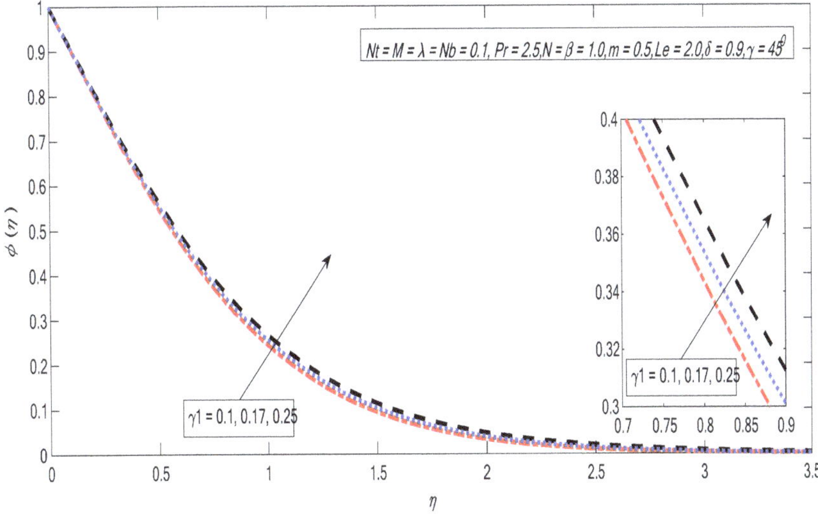

Figure 14. Variations in concentration profile for several values of γ_1.

Figure 15 represents that the temperature profile drops by improving the parameter Pr. This is because the bigger values of Pr cause improvement in viscosity and decline in the thermal boundary layer thickness. Figure 16 shows the result of Lewis number Le on the concentration profile. The boundary layer viscosity reduces by enhancing the values of Lewis number Le.

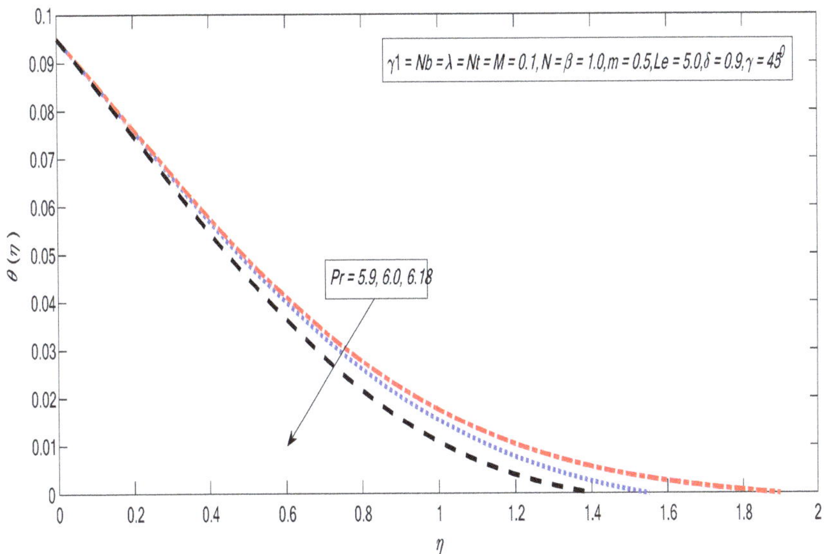

Figure 15. Variations in temperature profile for several values of Pr.

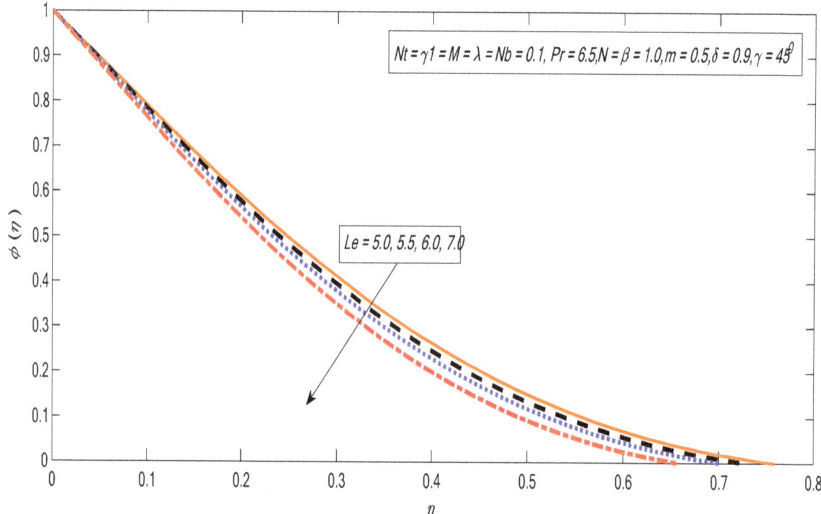

Figure 16. Variations in concentration profile for several values of *Le*.

Figures 17 and 18 show the effect of Brownian motion on the temperature and concentration profiles, respectively. The temperature profile upturns with improving Nb; on the other hand, concentration distribution has the opposite impact. Physically, the boundary layer heats up due to the development in Brownian motion, which accelerates the nanoparticles from the extending sheet to the stationary fluid. Therefore, the concentration of nanoparticles reduces. Figures 19 and 20 reveal the impact of Nt on temperature and concentration profile for altered values. It is observed that both temperature and concentration contours are directly proportional to the Nt.

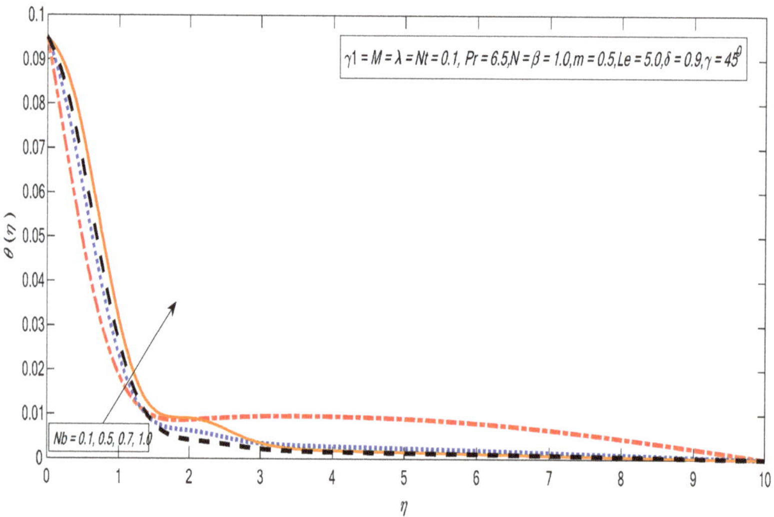

Figure 17. Variations in temperature profile for several values of *Nb*.

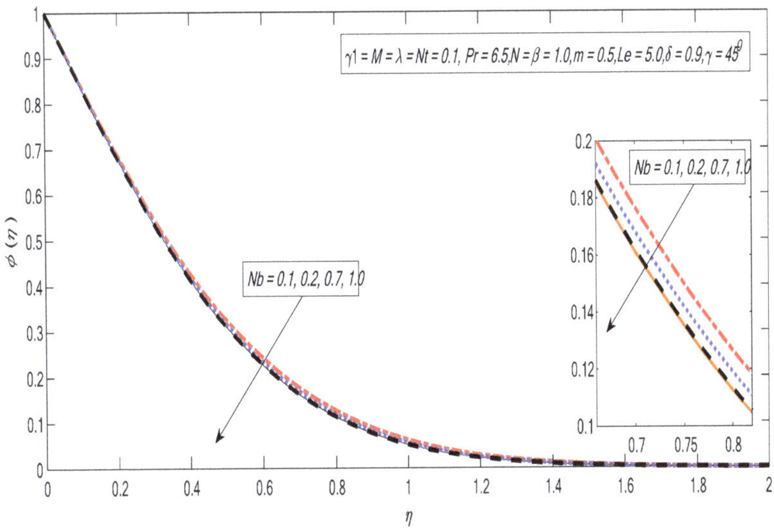

Figure 18. Variations in concentration profile for several values of Nb.

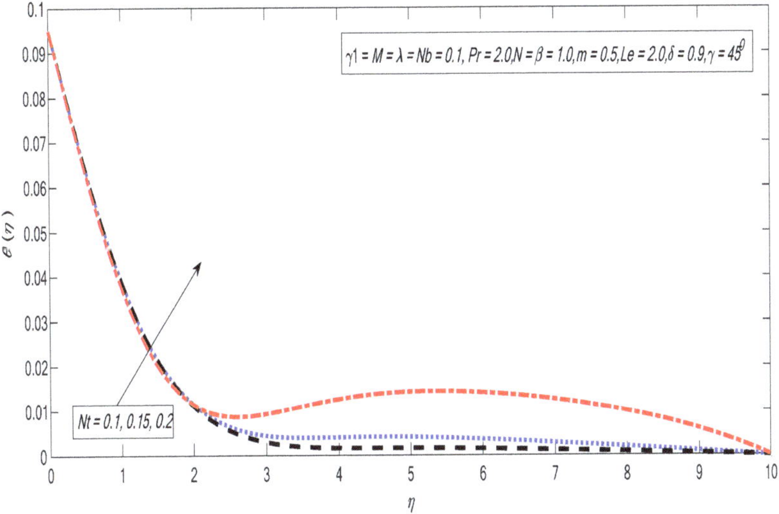

Figure 19. Variations in temperature profile for several values of Nt.

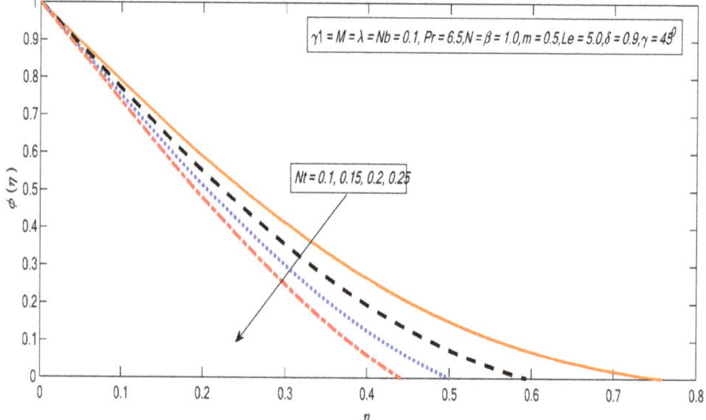

Figure 20. Variations in concentration profile for several values of Nt.

4. Conclusions

In the article under study, we investigate the heat and mass transfer of Casson nanofluid flow over an inclined sheet, with convective boundaries and thermal radiation effects taken in account. The numerical results are elucidated with the Keller-box method. We found an excellent agreement between the current outcomes and already published results. The core findings of the problem under concern are the following:

- The temperature profile increases near the boundary layer by improving the Biot number.
- The velocity and temperature profiles drop by improving the nonlinear power index.
- The heat exchange improved upon improving the radiation parameter.
- The velocity distribution retards by increasing the Casson parameter.
- The Nuselt number decreases by increasing the Casson parameter.
- The skin friction declines by improving the Biot number.
- The velocity profile shows an inverse relation with the inclination factor.

Author Contributions: M.I.A. and K.R. formulated the problem. M.M.; S.O.A. and P.T. elucidated the numerical results. K.S.N. and I.K. plotted and discussed the results.

Funding: This research received no external funding.

Acknowledgments: The authors would like to thank Deanship of Scientific Research, Majmaah University for supporting this work.

Conflicts of Interest: The authors declare that they have no conflict of interest.

References

1. Anwar, M.I.; Tanveer, N.; Salleh, M.Z.; Shafie, S. Diffusive effects on hydrodynamic Casson nanofluid boundary layer flow over a stretching surface. In *Journal of Physics: Conference Series*; IOP Publishing: Bristol, UK, 2017; Volume 890, p. 012047.
2. Afify, A.A. The influence of slip boundary condition on Casson nanofluid flow over a stretching sheet in the presence of viscous dissipation and chemical reaction. *Math. Probl. Eng.* **2017**, *2017*, 3804751. [CrossRef]
3. Souayeh, B.; Reddy, M.G.; Sreenivasulu, P.; Poornima, T.; Rahimi-Gorji, M.; Alarifi, I.M. Comparative analysis on non-linear radiative heat transfer on MHD Casson nanofluid past a thin needle. *J. Mol. Liq.* **2019**, *284*, 163–174. [CrossRef]
4. Rashidi, S.; Esfahani, J.; Ellahi, R. Convective heat transfer and particle motion in an obstructed duct with two side by side obstacles by means of DPM model. *Appl. Sci.* **2017**, *7*, 431. [CrossRef]
5. Bhatti, M.M.; Zeeshan, A.; Ellahi, R. Electromagnetohydrodynamic (EMHD) peristaltic flow of solid particles in a third-grade fluid with heat transfer. *Mech. Ind.* **2017**, *18*, 314. [CrossRef]

6. Ellahi, R.; Zeeshan, A.; Hussain, F.; Abbas, T. Study of shiny film coating on multi-fluid flows of a rotating disk suspended with nano-sized silver and gold particles: A comparative analysis. *Coatings* **2018**, *8*, 422. [CrossRef]
7. Anwar, M.I.; Shafie, S.; Hayat, T.; Shehzad, S.A.; Salleh, M.Z. Numerical study for MHD stagnation-point flow of a micropolar nanofluid towards a stretching sheet. *J. Braz. Soc. Mech. Sci. Eng.* **2017**, *39*, 89–100. [CrossRef]
8. Mustafa, M.; Khan, J.A. Model for flow of Casson nanofluid past a non-linearly stretching sheet considering magnetic field effects. *AIP Adv.* **2015**, *5*, 077148. [CrossRef]
9. Anwar, M.I.; Shafie, S.; Khan, I.; Ali, A.; Salleh, M.Z. Dufour and Soret effects on free convection flow of Nanofluid past a power law stretching sheet. *Int. J. Appl. Maths Stats* **2013**, *43*.
10. Anwar, M.I.; Khan, I.; Sharidan, S.; Salleh, M.Z. Conjugate effects of heat and mass transfer of nanofluids over a nonlinear stretching sheet. *Int. J. Phys. Sci.* **2012**, *7*, 4081–4092. [CrossRef]
11. Qing, J.; Bhatti, M.; Abbas, M.; Rashidi, M.; Ali, M. Entropy generation on MHD Casson nanofluid flow over a porous stretching/shrinking surface. *Entropy* **2016**, *18*, 123. [CrossRef]
12. Abbas, T.; Bhatti, M.M.; Ayub, M. Aiding and opposing of mixed convection Casson nanofluid flow with chemical reactions through a porous Riga plate. *Proc. Inst. Mech. Eng. Part E J. Process. Mech. Eng.* **2018**, *232*, 519–527. [CrossRef]
13. Choi, S.U.S.; Singer, D.A.; Wang, H.P. Developments and applications of non-Newtonian flows. *ASME FED* **1995**, *66*, 99–105. [CrossRef]
14. Asma, K.; Khan, I.; Arshad, K.; Sharidan, S.; Tlili, I. Case study of MHD blood flow in a porous medium with CNTS and thermal analysis. *Case Stud. Therm. Eng.* **2018**, *12*, 374–380.
15. Suriyakumar, P.; Devi, S.A. Effects of Suction and Internal Heat Generation on Hydromagnetic Mixed Convective Nanofluid Flow over an Inclined Stretching Plate. *Eur. J. Adv. Eng. Technol.* **2015**, *2*, 51–58.
16. Khan, M.; Shahid, A.; Malik, M.Y.; Salahuddin, T. Thermal and concentration diffusion in Jeffery nanofluid flow over an inclined stretching sheet: A generalized Fourier's and Fick's perspective. *J. Mol. Liq.* **2018**, *251*, 7–14. [CrossRef]
17. Thumma, T.; Beg, O.A.; Kadir, A. Numerical study of heat source/sink effects on dissipative magnetic nanofluid flow from a non-linear inclined stretching/shrinking sheet. *J. Mol. Liq.* **2017**, *232*, 159–173. [CrossRef]
18. Prakash, J.; Tripathi, D.; Triwari, A.K.; Sait, S.M.; Ellahi, R. Peristaltic Pumping of Nanofluids through a Tapered Channel in a Porous Environment: Applications in Blood Flow. *Symmetry* **2019**, *11*, 868. [CrossRef]
19. Zeeshan, A.; Shehzad, N.; Abbas, T.; Ellahi, R. Effects of radiative electro-magnetohydrodynamics diminishing internal energy of pressure-driven flow of titanium dioxide-water nanofluid due to entropy generation. *Entropy* **2019**, *21*, 236. [CrossRef]
20. Shehzad, N.; Zeeshan, A.; Ellahi, R.; Rashidi, S. Modelling study on internal energy loss due to entropy generation for non-darcy poiseuille flow of silver-water nanofluid: An application of purification. *Entropy* **2018**, *20*, 851. [CrossRef]
21. Hussain, F.; Ellahi, R.; Zeeshan, A. Mathematical models of electro-magnetohydrodynamic multiphase flows synthesis with nano-sized hafnium particles. *Appl. Sci.* **2018**, *8*, 275. [CrossRef]
22. Ellahi, R.; Zeeshan, A.; Hussain, F.; Abbas, T. Thermally Charged MHD Bi-phase Flow Coatings with non-Newtonian Nanofluid and Hafnium Particles along Slippery Walls. *Coatings* **2019**, *9*, 300. [CrossRef]
23. Rashad, A. Unsteady nanofluid flow over an inclined stretching surface with convective boundary condition and anisotropic slip impact. *Int. J. Heat Technol.* **2017**, *35*, 82–90. [CrossRef]
24. Govindarajan, A. Radiative fluid flow of a nanofluid over an inclined plate with non-uniform surface temperature. In *Journal of Physics: Conference Series*; IOP Publishing: Bristol, UK, 2018; Volume 1000, p. 12173.
25. Hatami, M.; Jing, D.; Yousif, M.A. Three-dimensional analysis of condensation nanofluid film on an inclined rotating disk by efficient analytical methods. *Arab J. Basic Appl. Sci.* **2018**, *25*, 28–37. [CrossRef]
26. Cimpean, D.S.; Pop, I. Fully developed mixed convection flow of a nanofluid through an inclined channel filled with a porous medium. *Int. J. Heat Mass Transf.* **2012**, *55*, 907–914. [CrossRef]
27. Reddy, P.B.A. Magnetohydrodynamic flow of a Casson fluid over an exponentially inclined permeable stretching surface with thermal radiation and chemical reaction. *Ain Shams Eng. J.* **2016**, *7*, 593–602. [CrossRef]

28. Hakeem, A.A.; Renuka, P.; Ganesh, N.V.; Kalaivanan, R.; Ganga, B. Influence of inclined Lorentz forces on boundary layer flow of Casson fluid over an impermeable stretching sheet with heat transfer. *J. Magn. Magn. Mater.* **2016**, *401*, 354–361. [CrossRef]
29. Rawi, N.A.; Ilias, M.R.; Lim, Y.J.; Isa, Z.M.; Shafie, S. Unsteady mixed convection flow of Casson fluid past an inclined stretching sheet in the presence of nanoparticles. In *Journal of Physics: Conference Series*; IOP Publishing: Bristol, UK, 2017; Volume 890, p. 012048.
30. Jain, S.; Parmar, A. Multiple slip effects on inclined MHD Casson fluid flow over a permeable stretching surface and a melting surface. *Int. J. Heat Technol.* **2018**, *36*, 585–594. [CrossRef]
31. Ali, M.; Alim, M.A.; Alam, M.S. Similarity solution of heat and mass transfer flow over an inclined stretching sheet with viscous dissipation and constant heat flux in presence of magnetic field. *Procedia Eng.* **2015**, *105*, 557–569. [CrossRef]
32. Ellahi, R.; Zeeshan, A.; Hussain, F.; Asadollahi, A. Peristaltic blood flow of couple stress fluid suspended with nanoparticles under the influence of chemical reaction and activation energy. *Symmetry* **2019**, *11*, 276. [CrossRef]
33. Bohra, S. Heat and mass transfer over a three-dimensional inclined non-linear stretching sheet with convective boundary conditions. *Indian J. Pure Appl. Phys. (IJPAP)* **2017**, *55*, 847–856.
34. Khan, M.; Sardar, H.; Gulzar, M.M.; Alshomrani, A.S. On multiple solutions of non-Newtonian Carreau fluid flow over an inclined shrinking sheet. *Results Phys.* **2018**, *8*, 926–932. [CrossRef]
35. Alam, M.S.; Islam, M.R.; Ali, M.; Alim, M.A.; Alam, M.M. Magnetohydrodynamic boundary layer flow of non-Newtonian fluid and combined heat and mass transfer about an inclined stretching sheet. *Open J. Appl. Sci.* **2015**, *5*, 279. [CrossRef]
36. Rafique, K.; Anwar, M.I.; Misiran, M. Numerical Study on Micropolar Nanofluid Flow over an Inclined Surface by Means of Keller-Box. *Asian J. Probab. Stat.* **2019**, 1–21. [CrossRef]
37. Bognár, G.; Gombkötő, I.; Hriczó, K. Non-Newtonian fluid flow down an inclined plane. In Proceedings of the 9th ASME/WSEAS International Conference on Fluid Mechanics and Aerodynamics, Florence, Italy, August 2011; pp. 23–25.
38. Reddy, V.R.; REDDY, M.S.; Nagendra, N.; Rao, A.S.; Reddy, M.S. Radiation Effect on Boundary Layer Flow of a Non Newtonian Jeffrey Fluid Past an Inclined Vertical Plate. *i-Manag. J. Math.* **2017**, *6*, 34.
39. Rafique, K.; Anwar, M.I.; Misiran, M. Keller-box Study on Casson Nano Fluid Flow over a Slanted Permeable Surface with Chemical Reaction. *Asian Res. J. Math.* **2019**, *14*, 1–17. [CrossRef]
40. Ellahi, R.; Zeeshan, A.; Hussain, F.; Abbas, T. Two-Phase Couette Flow of Couple Stress Fluid with Temperature Dependent Viscosity Thermally Affected by Magnetized Moving Surface. *Symmetry* **2019**, *11*, 647. [CrossRef]
41. Benos, L.T.; Mahabaleshwar, U.S.; Sakanaka, P.H.; Sarris, I.E. Thermal analysis of the unsteady sheet stretching subject to slip and magnetohydrodynamic effects. *Therm. Sci. Eng. Prog.* **2019**, *13*, 100367. [CrossRef]
42. Mastroberardino, A.; Mahabaleshwar, U.S. Mixed convection in viscoelastic flow due to a stretching sheet in a porous medium. *J. Porous Media* **2013**, *16*, 483–500. [CrossRef]
43. Ghadikolaei, S.S.; Hosseinzadeh, K.; Ganji, D.D.; Jafari, B. Nonlinear thermal radiation effect on magneto Casson nanofluid flow with Joule heating effect over an inclined porous stretching sheet. *Case Stud. Therm. Eng.* **2018**, *12*, 176–187. [CrossRef]
44. Saidulu, N.; Gangaiah, T.; Lakshmi, A.V. Radiation Effect on Mhd Flow of a Tangent Hyperbolic Nanofluid over an Inclined Exponentially Stretching Sheet. *Int. J. Fluid Mech. Res.* **2019**, *46*. [CrossRef]
45. Khan, W.A.; Pop, I. Boundary-layer flow of a nanofluid past a stretching sheet. *Int. J. Heat Mass Transf.* **2010**, *53*, 2477–2483. [CrossRef]
46. Hussanan, A.; Salleh, M.Z.; Alkasasbeh, H.T.; Khan, I. MHD flow and heat transfer in a Casson fluid over a nonlinearly stretching sheet with Newtonian heating. *Heat Transf. Res.* **2018**, *49*. [CrossRef]
47. Ullah, A.; Shah, Z.; Kumam, P.; Ayaz, M.; Islam, S.; Jameel, M. Viscoelastic MHD Nanofluid Thin Film Flow over an Unsteady Vertical Stretching Sheet with Entropy Generation. *Processes* **2019**, *7*, 262. [CrossRef]
48. Oyelakin, I.S.; Mondal, S.; Sibanda, P. Unsteady Casson nanofluid flow over a stretching sheet with thermal radiation, convective and slip boundary conditions. *Alex. Eng. J.* **2016**, *55*, 1025–1035. [CrossRef]

Article

Numerical Study for Magnetohydrodynamic Flow of Nanofluid Due to a Rotating Disk with Binary Chemical Reaction and Arrhenius Activation Energy

Mir Asma [1], W.A.M. Othman [1,*], Taseer Muhammad [2], Fouad Mallawi [3] and B.R. Wong [1]

1 Institute of Mathematical Sciences, Faculty of Science, University of Malaya, Kuala Lumpur 50603, Malaysia; miraszaka@gmail.com (M.A.); bernardr@um.edu.my (B.R.W.)

2 Department of Mathematics, Government College Women University, Sialkot 51310, Pakistan; taseer_qau@yahoo.com

3 Department of Mathematics, Faculty of Science, King Abdulaziz University, Jeddah 21589, Saudi Arabia; fmallawi@kau.edu.sa

* Correspondence: wanainun@um.edu.my

Received: 5 August 2019; Accepted: 1 October 2019; Published: 14 October 2019

Abstract: This article examines magnetohydrodynamic 3D nanofluid flow due to a rotating disk subject to Arrhenius activation energy and heat generation/absorption. Flow is created due to a rotating disk. Velocity, temperature and concentration slips at the surface of the rotating disk are considered. Effects of thermophoresis and Brownian motion are also accounted. The nonlinear expressions have been deduced by transformation procedure. Shooting technique is used to construct the numerical solution of governing system. Plots are organized just to investigate how velocities, temperature and concentration are influenced by various emerging flow parameters. Skin-friction Local Nusselt and Sherwood numbers are also plotted and analyzed. In addition, a symmetry is noticed for both components of velocity when Hartman number enhances.

Keywords: rotating disk; Arrhenius activation energy; nanoparticles; binary chemical reaction; MHD; heat generation/absorption; slip effects; numerical solution

---

1. Introduction

Low thermal efficiency of working liquids is a principle issue for a few heat transport components in the designing of applications. Therefore a few scientists are occupied with the request to build up an imaginative route for development of thermal productivity of working liquids. Different components have been proposed by specialists to improve the thermal productivity of liquids. Therefore, the inclusion of nanomaterial in working liquid termed as nanofluid is very alluring component. Recent examinations on nanofluid have uncovered that working fluid has various highlights with nanomaterial blend. This is on the grounds that the thermal proficiency of working fluid is weaker than nanofluid thermal productivity. Nanofluid is a recently perceived group of liquids containing working liquid with the particles of nano measure. Such nanomaterials are employed in MHD control generators, oil stores, cooling of atomic reactors, malignancy treatment, vehicle transformer and several others [1–5]. The word nanofluid was first used by Choi [6] to explain the thermal conductivity of ordinary liquids. From the perspective of exploring how thermal conductivity is expanded, various examinations are introduced by him. Further attempts on nanofluids can be cited through investigations [7–20].

Analysts are presently much occupied by exploring fluid flow via rotating disk. It is because of its numerous applications in various fields of technology, for example, design branches and aeronautical science such as gem development forms, electronic gadgets, pivoting hardware, PC stockpiling gadgets, thermal power producing systems, gas turbine rotors, air cleaning machines, restorative gear and several others [21–23]. Von Karman [24] provided pioneering work with fluid flow via rotating disk. He examined the subsequent issue diagnostically. Cochran [25] proved an asymptotic answer for the von Karman issue. Millsaps and Pohlhansen [26] examined the issue of heat transfer for the isothermal plate. Ackroyd [27] thought about suction/infusion impacts in the Von Karman issue and he created a solution by exponentially rotting coefficients. Miclavcic and Wang [28] broadened the Von Karman issue for circumstances where a rotating disk concedes partial slip attributes. Attia [29] examined fluid flow because of a rotating disk inundated in a permeable space by using the Wrench Nicolson technique. Flow of viscous fluid by permeable disk subject to pivoting casing and heat/mass exchange was analyzed by Turkyilmazoglu and Senel [30]. They processed numeric consequences of governing flow issue. Rashidi et al. [31] analyzed impacts of entropy generation in MHD flow of viscous fluid by rotating disk. Hatami et al. [32] talked about laminar flow of nanofluids instigated by turning contracting disk. Mustafa et al. [33] investigated the flow of nanoliquid initiated by a rotating disk. They inferred that uniform extension of a disk is a significant factor for decreasing boundary-layer thickness. Sheikholeslami et al. [34] accounted for nanofluid flow incited by a slanted rotating plate. Hayat et al. [35] examined flow by a rotating disk through a magnetic field, slip and nanoparticle impact. Flow of MHD nanoliquid by rotating disk subject to slip was explored by Mustafa [36]. Darcy Forchheimer flow of carbon nanotubes incited by a rotating disk was examined by Hayat et al. [37]. Further relevant attempts regarding rotating disks can be seen through investigations [38–40].

Propelled by the above articles, the goal here is to look at the combined impacts of Arrhenius activation energy and binary chemical reactions in hydromagnetic 3D flow of nanofluid by rotating disk with heat generation/absorption and slip impacts. The random movement and thermophoretic dispersion phenomena occur because of the nanoparticles. Velocity, thermal and concentration slips are considered. The governing system is solved numerically by shooting procedure. Velocities, temperature, concentration and local Sherwood and Nusselt numbers are additionally discussed through curves.

2. Statement

We analyze MHD steady three-dimensional flow of nanoliquid by rotating disk with thermal generation/absorption and slip impacts. Arrhenius activation energy and binary chemical reaction impacts are additionally present. Disk at $z = 0$ pivots with constant angular velocity Ω. Brownian dispersion and thermophoretic impacts are also present. Magnetic field of strength B_0 acts in $z-$direction (see Figure 1). The velocity components (u, v, w) are in the directions of expanding (r, φ, z) respectively. Resulting boundary-layer expressions are

$$\frac{\partial u}{\partial r} + \frac{u}{r} + \frac{\partial w}{\partial z} = 0, \tag{1}$$

$$u\frac{\partial u}{\partial r} - \frac{v^2}{r} + w\frac{\partial u}{\partial z} = \nu\left(\frac{\partial^2 u}{\partial r^2} + \frac{1}{r}\frac{\partial u}{\partial r} - \frac{u}{r^2} + \frac{\partial^2 u}{\partial z^2}\right) - \frac{\sigma B_0^2}{\rho_f}u, \tag{2}$$

$$u\frac{\partial v}{\partial r} + \frac{uv}{r} + w\frac{\partial v}{\partial z} = \nu\left(\frac{\partial^2 v}{\partial r^2} + \frac{1}{r}\frac{\partial v}{\partial r} - \frac{v}{r^2} + \frac{\partial^2 v}{\partial z^2}\right) - \frac{\sigma B_0^2}{\rho_f}v, \tag{3}$$

$$u\frac{\partial w}{\partial r} + w\frac{\partial w}{\partial z} = \nu\left(\frac{\partial^2 w}{\partial r^2} + \frac{1}{r}\frac{\partial w}{\partial r} + \frac{\partial^2 w}{\partial z^2}\right), \tag{4}$$

$$\begin{aligned} u\frac{\partial T}{\partial r} + w\frac{\partial T}{\partial z} =& \alpha_m\left(\frac{\partial^2 T}{\partial r^2} + \frac{1}{r}\frac{\partial T}{\partial r} + \frac{\partial^2 T}{\partial z^2}\right) + \frac{Q}{(\rho c)_f}(T - T_\infty) \\ &+ \frac{(\rho c)_p}{(\rho c)_f}\left(D_B\left(\frac{\partial T}{\partial r}\frac{\partial C}{\partial r} + \frac{\partial T}{\partial z}\frac{\partial C}{\partial z}\right) + \frac{D_T}{T_\infty}\left(\left(\frac{\partial T}{\partial r}\right)^2 + \left(\frac{\partial T}{\partial z}\right)^2\right)\right), \end{aligned} \tag{5}$$

$$u\frac{\partial C}{\partial r}+w\frac{\partial C}{\partial z}=D_B\left(\frac{\partial^2 C}{\partial r^2}+\frac{1}{r}\frac{\partial C}{\partial r}+\frac{\partial^2 C}{\partial z^2}\right)+\frac{D_T}{T_\infty}\left(\frac{\partial^2 T}{\partial r^2}+\frac{1}{r}\frac{\partial T}{\partial r}+\frac{\partial^2 T}{\partial z^2}\right)-k_r^2\left(C-C_\infty\right)\left(\frac{T}{T_\infty}\right)^n\exp\left(-\frac{E_a}{\kappa T}\right), \tag{6}$$

$$u=L_1\frac{\partial u}{\partial z},\ v=r\Omega+L_1\frac{\partial v}{\partial z},\ w=0,\ T=T_w+L_2\frac{\partial T}{\partial z},\ C=C_w+L_3\frac{\partial C}{\partial z}\ \text{at}\ z=0, \tag{7}$$

$$u\to 0,\quad v\to 0,\quad T\to T_\infty,\quad C\to C_\infty\quad \text{when}\ z\to\infty. \tag{8}$$

Here u, v and w represent velocities in directions of r, φ and z while ρ_f, $\nu\left(=\mu/\rho_f\right)$ and μ show density, kinematic and dynamic viscosities, respectively, L_1 the velocity slip factor, $(\rho c)_p$ the effective heat capacity of nanoparticles, E_a the activation energy, $(\rho c)_f$ heat capacity of liquid, L_2 the thermal slip factor, σ the electrical conductivity, C the concentration, n the fitted rate constant, C_∞ the ambient concentration, D_T the thermophoretic factor, $\alpha_m = k/(\rho c)_f$ and k the thermal diffusivity and thermal conductivity respectively, T the fluid temperature, k_r the reaction rate, D_B the Brownian factor, L_3 the concentration slip factor, Q the heat generation/absorption factor, κ the Boltzmann constant and T_∞ the ambient temperature. Selecting

$$\left.\begin{array}{c} u=r\Omega f'(\zeta),\ w=-(2\Omega\nu)^{1/2}f(\zeta),\ v=r\Omega g(\zeta), \\ \phi(\zeta)=\frac{C-C_\infty}{C_w-C_\infty},\ \zeta=\left(\frac{2\Omega}{\nu}\right)^{1/2}z,\ \theta(\zeta)=\frac{T-T_\infty}{T_w-T_\infty}. \end{array}\right\} \tag{9}$$

Continuity Equation (1) is trivially verified while Equations (2)–(8) yield

$$2f'''+2ff''-f'^2+g^2-(Ha)^2f'=0, \tag{10}$$

$$2g''+2fg'-2f'g-(Ha)^2g=0, \tag{11}$$

$$\frac{1}{\Pr}\theta''+f\theta'+N_b\theta'\phi'+N_t\theta'^2+\delta_1\theta=0, \tag{12}$$

$$\frac{1}{Sc}\phi''+f\phi'+\frac{1}{Sc}\frac{N_t}{N_b}\theta''-\sigma\left(1+\delta\theta\right)^n\phi\exp\left(-\frac{E}{1+\delta\theta}\right)=0, \tag{13}$$

$$f(0)=0,\ f'(0)=\alpha f''(0),\ g(0)=1+\alpha g'(0),\ \theta(0)=1+\beta\theta'(0),\ \phi(0)=1+\gamma\phi'(0), \tag{14}$$

$$f'(\infty)\to 0,\ g(\infty)\to 0,\ \theta(\infty)\to 0,\ \phi(\infty)\to 0. \tag{15}$$

Here N_t stands for the thermophoresis parameter, α for velocity slip parameter, Ha for Hartman number, σ for chemical reaction parameter, N_b for Brownian parameter, β for thermal slip parameter, δ for temperature difference parameter, Pr for Prandtl number, γ for concentration slip parameter, δ_1 for heat absorption/generation parameter, Sc for Schmidt number and E for nondimensional activation energy. Nondimensional variables are defined by

$$\left.\begin{array}{c} (Ha)^2=\frac{\sigma B_0^2}{\Omega\rho_f},\ \Pr=\frac{\nu}{\alpha_m},\ \alpha=L_1\sqrt{\frac{2\Omega}{\nu}},\ \gamma=L_3\sqrt{\frac{2\Omega}{\nu}}, \\ N_t=\frac{(\rho c)_p D_T(T_w-T_\infty)}{(\rho c)_f\nu T_\infty},\ Sc=\frac{\nu}{D_B},\ N_b=\frac{(\rho c)_p D_B(C_w-C_\infty)}{(\rho c)_f\nu},\ \delta_1=\frac{Q}{2\Omega(\rho c)_f}, \\ \beta=L_2\sqrt{\frac{2\Omega}{\nu}},\ \sigma=\frac{k_r^2}{\Omega},\ \delta=\frac{T_f-T_\infty}{T_\infty},\ E=\frac{E_a}{\kappa T_\infty}. \end{array}\right\} \tag{16}$$

The coefficients of skin-friction and local Nusselt and Sherwood expressions are

$$\left.\begin{array}{c} \mathrm{Re}_r^{1/2}C_f=f''(0),\ \mathrm{Re}_r^{1/2}C_g=g'(0), \\ \mathrm{Re}_r^{-1/2}Nu=-\theta'(0),\ \mathrm{Re}_r^{-1/2}Sh=-\phi'(0), \end{array}\right\} \tag{17}$$

where $\mathrm{Re}_r=2(\Omega r)r/\nu$ represents the local rotational Reynolds number.

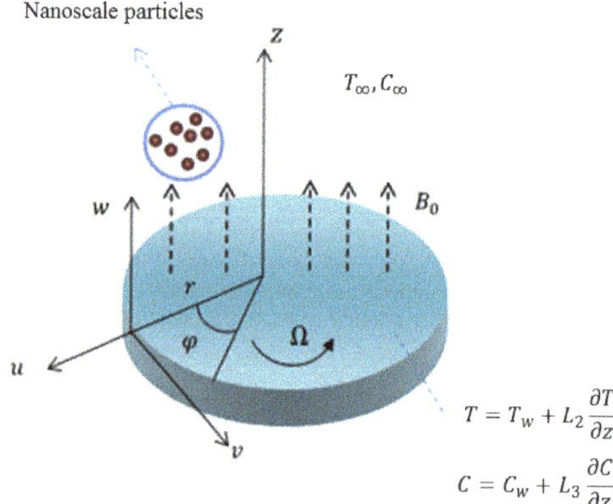

Figure 1. Flow configuration.

3. Solution Methodology

By using appropriate boundary conditions on a system of equations, a numerical solution is provided considering NDSolve in Mathematica. Shooting technique is used via NDSolve. This technique is very helpful in the case of small step-sizes featuring negligible error. Consequently, both r and z varied uniformly by a step-size of 0.01.

4. Results and Discussion

This segment outlines the commitment of various relevant parameters including Prandtl number Pr, Hartman number Ha, thermophoresis parameter N_t, chemical reaction parameter σ, velocity slip parameter α, Schmidt number Sc, temperature difference parameter δ, Brownian parameter N_b, thermal slip parameter β, activation energy E, heat generation/absorption parameter δ_1, concentration slip parameter γ on velocities $f'(\zeta)$ and $g(\zeta)$, concentration $\phi(\zeta)$ and temperature $\theta(\zeta)$ distributions. Figure 2 demonstrates the variety in velocity field $f'(\zeta)$ for shifting Hartman parameter Ha. An addition in Hartman parameter Ha relates to bringing down velocity field $f'(\zeta)$. Here Ha=0 yields hydromagnetic flow circumstance and Ha=0 speaks to hydro-dynamic flow case. Figure 3 portrays adjustment in velocity field $f'(\zeta)$ for differing estimations of velocity slip parameter α. Velocity field and related layer are diminished for higher α. Figure 4 shows adjustment in velocity field $g(\zeta)$ for fluctuating Hartman parameter Ha. Here we examined that velocity field diminishes when Hartman parameter Ha increments. Figure 5 is portrayed to look at that how velocity field $g(\zeta)$ is influenced with variety of velocity slip α. For increasing estimations of α, velocity field $g(\zeta)$ indicates diminishing pattern. Figure 6 showcases the impact of Hartman parameter Ha on temperature $\theta(\zeta)$. Obviously, temperature and the related thermal layer are upgraded for increasing Ha. Effect of thermal slip β on temperature dissemination $\theta(\zeta)$ is delineated in Figure 7. Improvement in β depicts diminishing conduct for $\theta(\zeta)$ and the related thermal layer. Figure 8 shows how heat generation/absorption number δ_1 influences temperature dispersion $\theta(\zeta)$. Here $\delta_1 > 0$ portrays heat generation and $\delta_1 < 0$ for heat absorption. Both temperature $\theta(\zeta)$ and thermal layer are upgraded for increasing δ_1. Figure 9 introduces a variety in temperature field $\theta(\zeta)$ for differing Prandtl parameter Pr. Here $\theta(\zeta)$ is diminished through Pr. Proportion of momentum diffusivity to thermal diffusivity is termed as Prandtl parameter Pr. Higher estimations of Pr yield more fragile thermal diffusivity which compares to a reduction in thermal layer. Effect of N_t on temperature profile $\theta(\zeta)$ is depicted in Figure 10. Addition in N_t relates to stronger temperature field $\theta(\zeta)$ and more thermal layer. Figure 11 delineates variety in temperature field $\theta(\zeta)$ for unmistakable estimations of Brownian movement N_b. Physically, a sporadic movement of nanoparticles improves by expanding Brownian movement N_b because of which impact of particles happens. As a result, dynamic vitality is changed into warmth vitality which shows an upgrade in temperature profile and the related layer. Figure 12 demonstrates that how Hartman parameter Ha influences concentration $\phi(\zeta)$. By expanding Hartman parameter Ha, both concentration and concentration layers are improved. Figure 13 shows that concentration dispersion $\phi(\zeta)$ is weaker for bigger concentration slip. From Figure 14, we saw that bigger Schmidt parameter Sc demonstrates a rot in concentration field $\phi(\zeta)$. Schmidt parameter is conversely relative to Brownian diffusivity. Increasing Schmidt parameter Sc yields a more fragile Brownian diffusivity. This more fragile Brownian diffusivity prompts lower concentration field $\phi(\zeta)$. Figure 15 demonstrates that how thermophoresis N_t influences concentration profile $\phi(\zeta)$. By improving thermophoresis parameter N_t, the concentration field $\phi(\zeta)$ and related layer are expanded. Figure 16 portrays effect of Brownian movement N_b on concentration $\phi(\zeta)$. It is obviously observed that a more fragile concentration $\phi(\zeta)$ is produced by using higher Brownian movement parameter N_b. Figure 17 explains the impact of nondimensional activation energy E on concentration $\phi(\zeta)$. An improvement in activation energy E rots altered Arrhenius work $\left(\frac{T}{T_\infty}\right)^n \exp\left(-\frac{E_a}{\kappa T}\right)$. This inevitably builds up the generative synthetic response because of which concentration $\phi(\zeta)$ upgrades. Figure 18 introduces an improvement in compound response parameter σ shows a rot in concentration $\phi(\zeta)$ and its related layer. Figure 19 explains the impact of δ on $\phi(\zeta)$. Here $\phi(\zeta)$ is seen as a diminishing capacity of δ. Figure 20 depicts the concentration $\phi(\zeta)$ for evolving n. By improving n, infiltration profundity of $\phi(\zeta)$ closures become slenderer. Figures 21 and 22 display the effects of N_t and N_b on $\mathrm{Re}_r^{-1/2}\,Nu$. From these figures, it has been noticed that $\mathrm{Re}_r^{-1/2}\,Nu$ reduces for higher N_t and N_b. Features of N_t and N_b on $\mathrm{Re}_r^{-1/2}\,Sh$ are disclosed through Figures 23 and 24. Interestingly, $\mathrm{Re}_r^{-1/2}\,Sh$ is an increasing function of N_t while it is a decreasing function of N_b.

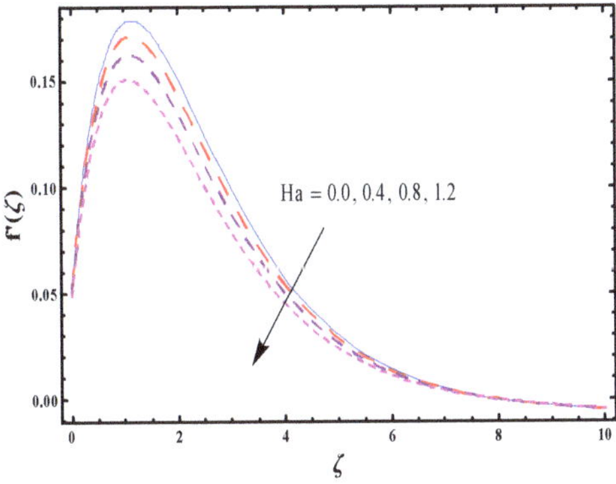

Figure 2. Sketch of $f'(\zeta)$ for Ha.

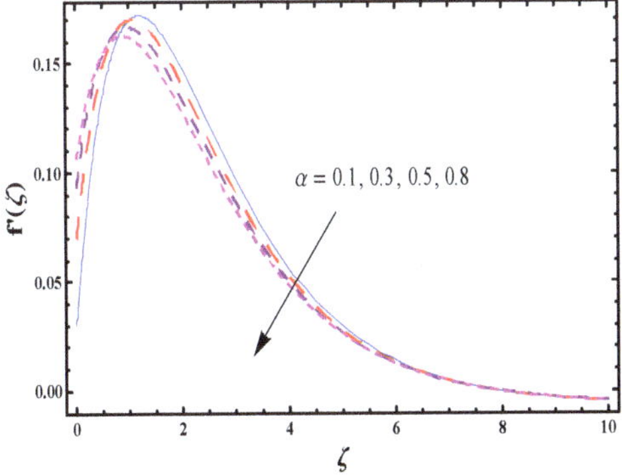

Figure 3. Sketch of $f'(\zeta)$ for α.

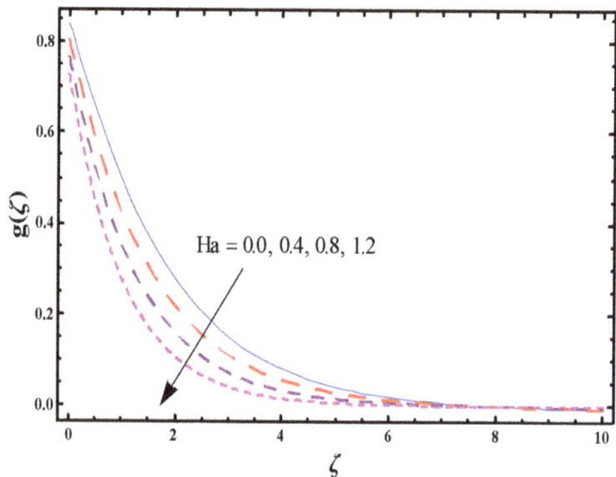

Figure 4. Sketch of $g(\zeta)$ for Ha.

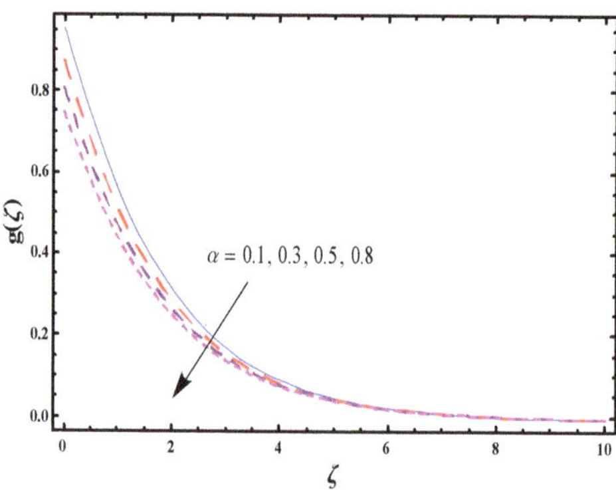

Figure 5. Sketch of $g(\zeta)$ for α.

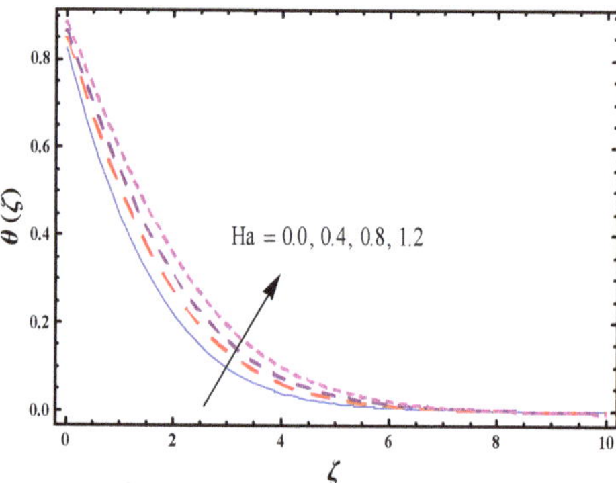

Figure 6. Sketch of $\theta(\zeta)$ for Ha.

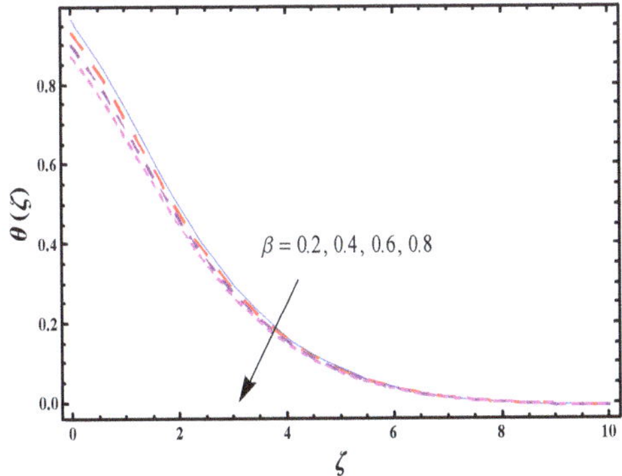

Figure 7. Sketch of $\theta(\zeta)$ for β.

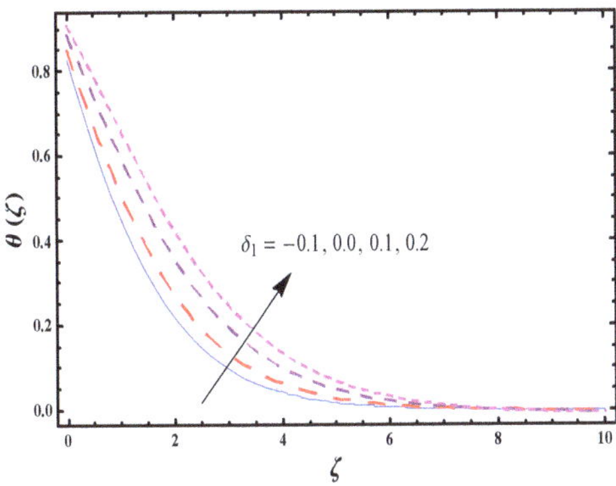

Figure 8. Sketch of $\theta(\zeta)$ for δ_1.

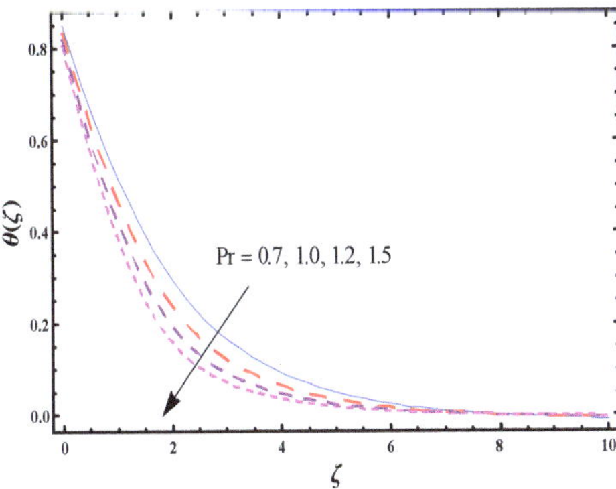

Figure 9. Sketch of $\theta(\zeta)$ for Pr.

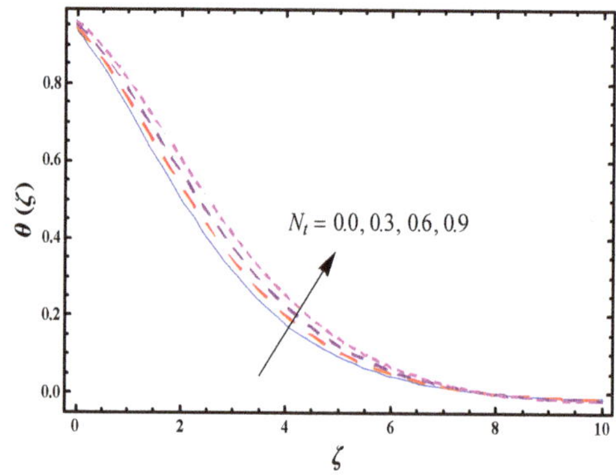

Figure 10. Sketch of $\theta(\zeta)$ for N_t.

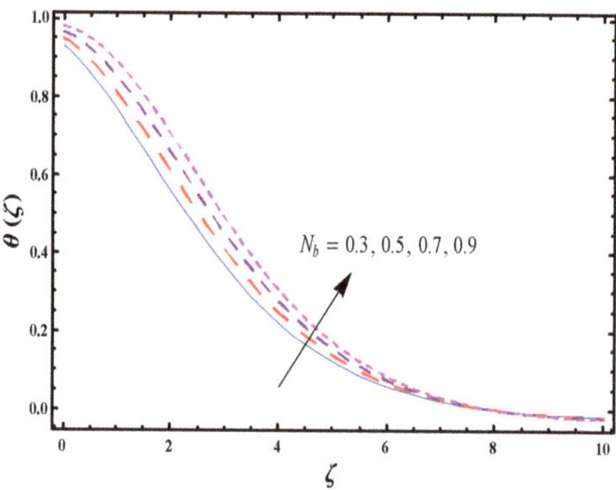

Figure 11. Sketch of $\theta(\zeta)$ for N_b.

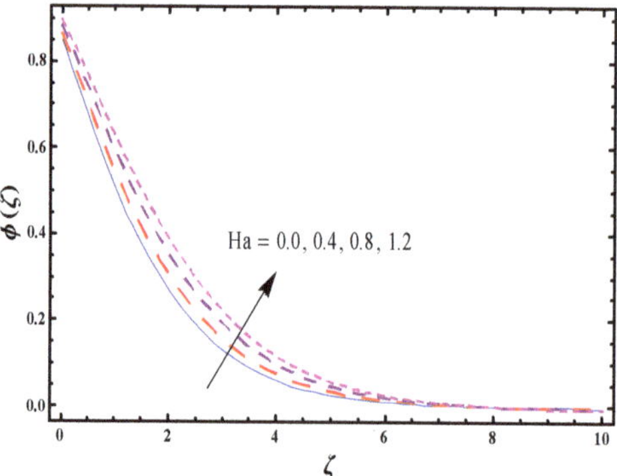

Figure 12. Sketch of $\phi(\zeta)$ for Ha.

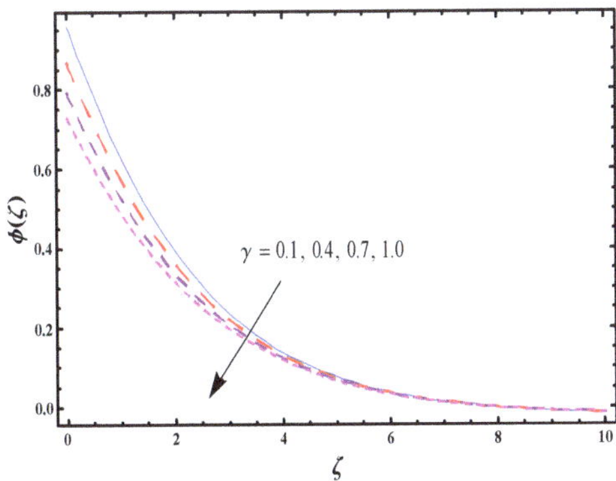

Figure 13. Sketch of $\phi(\zeta)$ for γ.

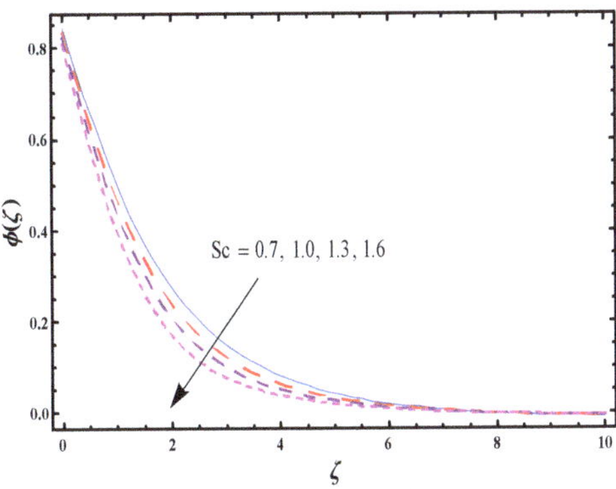

Figure 14. Sketch of $\phi(\zeta)$ for Sc.

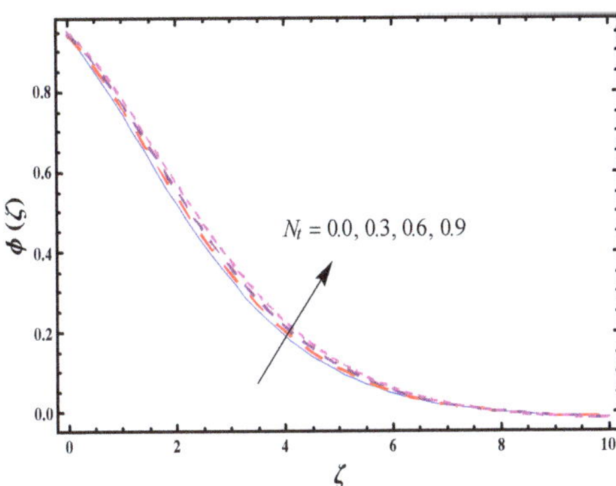

Figure 15. Sketch of $\phi(\zeta)$ for N_t.

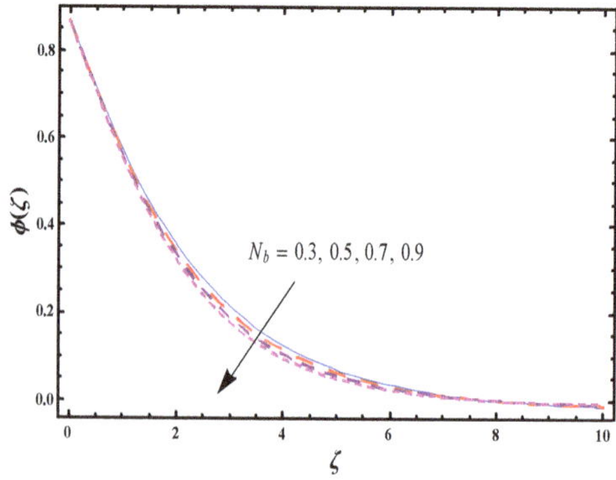

Figure 16. Sketch of $\phi(\zeta)$ for N_b.

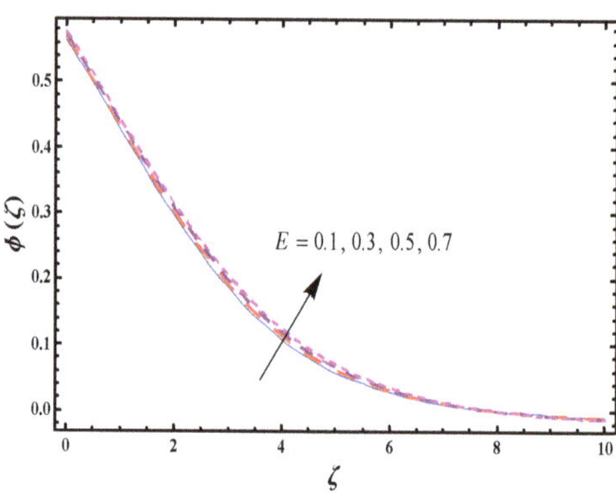

Figure 17. Sketch of $\phi(\zeta)$ for E.

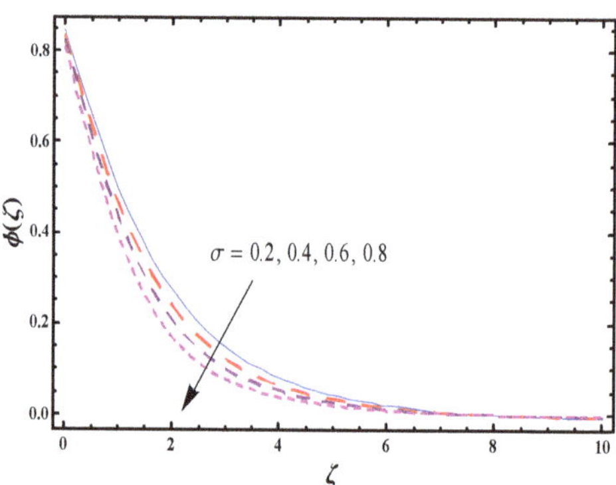

Figure 18. Sketch of $\phi(\zeta)$ for σ.

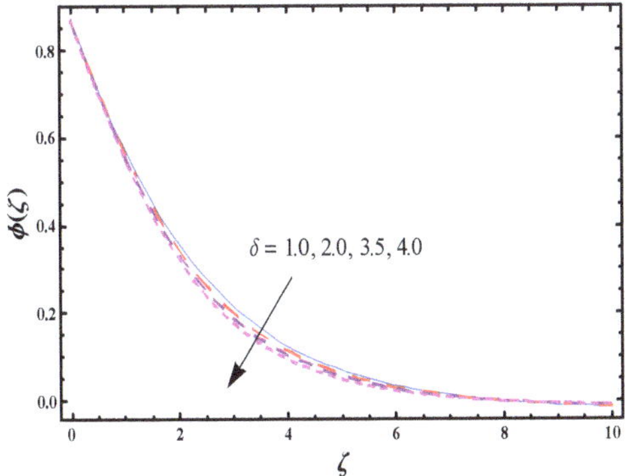

Figure 19. Sketch of $\phi(\zeta)$ for δ.

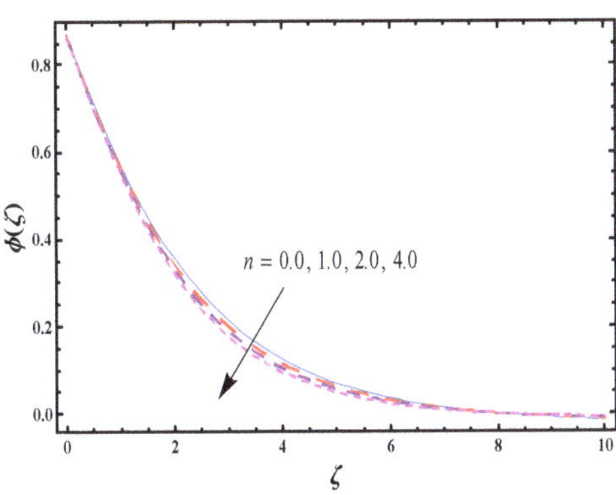

Figure 20. Sketch of $\phi(\zeta)$ for n.

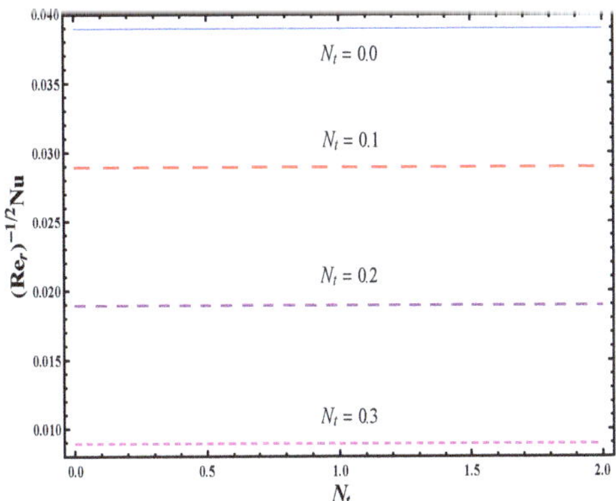

Figure 21. Sketch of $\mathrm{Re}_r^{-1/2}\,Nu$ for N_t.

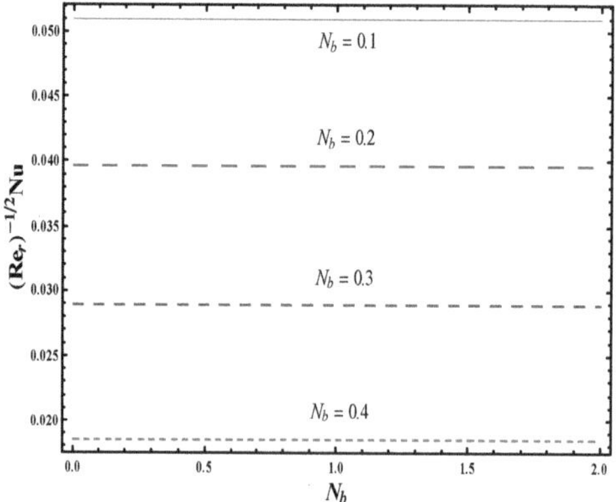

Figure 22. Sketch of $\mathrm{Re}_r^{-1/2}\, Nu$ for N_b.

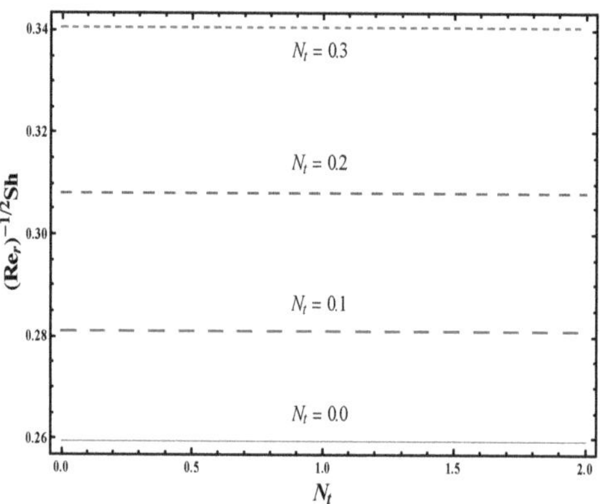

Figure 23. Sketch of $\mathrm{Re}_r^{-1/2}\, Sh$ for N_t.

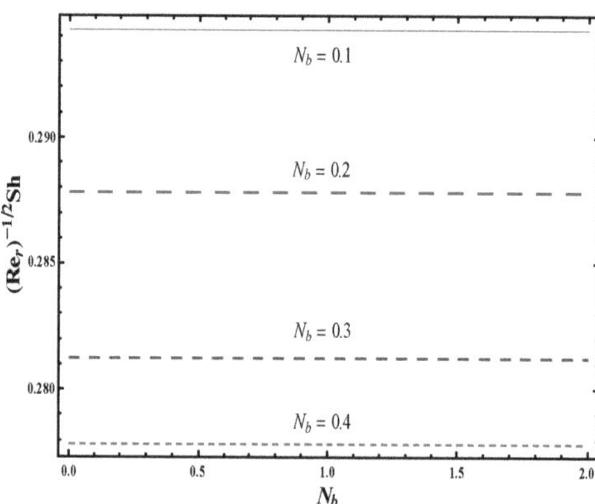

Figure 24. Sketch of $\mathrm{Re}_r^{-1/2}\, Sh$ for N_b.

5. Conclusions

Magnetohydrodynamic viscous nanoliquid 3D flow by rotating disk with heat absorption/generation, binary chemical reaction and Arrhenius activation energy is examined. Major results are given as follows:

- Larger velocity slip α and Hartman number Ha show decreasing trend for both velocities $f'(\zeta)$ and $g(\zeta)$.
- Both concentration and temperature depict increasing trend for increasing Ha.
- Higher Pr corresponds to weaker temperature while the reverse behavior is seen for δ_1.
- Stronger temperature distribution is seen for N_b and N_t.
- Higher γ exhibits a decreasing trend for the concentration field.
- Higher activation energy E shows stronger concentration $\phi(\zeta)$.
- Concentration $\phi(\zeta)$ depicts decreasing behavior for larger δ and σ.
- Both concentration $\phi(\zeta)$ is a decreasing factor of higher Sc.
- Concentration $\phi(\zeta)$ displays reverse behavior for N_b and N_t.

Author Contributions: All the authors have contributed equally to this manuscript.

Funding: This project was funded by the Deanship of Scientific Research (DSR), King Abdulaziz University, Jeddah, Saudi Arabia under grant no. (KEP-16-130-40).

Acknowledgments: This project was funded by the Deanship of Scientific Research (DSR), King Abdulaziz University, Jeddah, Saudi Arabia under grant no. (KEP-16-130-40). The authors, therefore, acknowledge with thanks DSR technical and financial support.

Conflicts of Interest: The authors declare no conflict of interest.

References

1. Hsiao, K.L. Nanofluid flow with multimedia physical features for conjugate mixed convection and radiation. *Comput. Fluids* **2014**, *104*, 1–8. [CrossRef]
2. Wen, B.; Corson, L.T.; Chini, G.P. Structure and stability of steady porous medium convection at large Rayleigh number. *J. Fluid Mech.* **2015**, *772*, 197–224. [CrossRef]
3. Hsiao, K.L. Stagnation electrical MHD nanofluid mixed convection with slip boundary on a stretching sheet. *Appl. Therm. Eng.* **2016**, *98*, 850–861. [CrossRef]
4. Hsiao, K.L. Micropolar nanofluid flow with MHD and viscous dissipation effects towards a stretching sheet with multimedia feature. *Int. J. Heat Mass Transf.* **2017**, *112*, 983–990. [CrossRef]
5. Wen, B.; Chang, K.W.; Hesse, M.A. Rayleigh-Darcy convection with hydrodynamic dispersion. *Phys. Rev. Fluids* **2018**, *3*, 123801. [CrossRef]
6. Choi, S.U.S. Enhancing thermal conductivity of fluids with nanoparticles. In Proceedings of the ASME International Mechanical Engineering Congress & Exposition, San Francisco, CA, USA, 12–17 November 1995; Volume 66, pp. 99–105.
7. Eastman, J.A.; Choi, S.U.S.; Li, S.; Yu, W.; Thompson, L.J. Anomalously increased effective thermal conductivities of ethylene glycol-based nanofluids containing copper nanoparticles. *Appl. Phys. Lett.* **2001**, *78*, 718–720. [CrossRef]
8. Buongiorno, J. Convective transport in nanofluids. *J. Heat Transf.* **2006**, *128*, 240–250. [CrossRef]
9. Tiwari, R.K.; Das, M.K. Heat transfer augmentation in a two-sided lid-driven differentially heated square cavity utilizing nanofluid. *Int. J. Heat Mass Transf.* **2007**, *50*, 2002–2018. [CrossRef]
10. Abu-Nada, E.; Oztop, H.F. Effects of inclination angle on natural convection in enclosures filled with Cu-water nanofluid. *Int. J. Heat Fluid Flow* **2009**, *30*, 669–678. [CrossRef]
11. Khan, J.A.; Mustafa, M.; Hayat, T.; Farooq, M.A.; Alsaedi, A.; Liao, S.J. On model for three-dimensional flow of nanofluid: An application to solar energy. *J. Mol. Liq.* **2014**, *194*, 41–47. [CrossRef]
12. Mansur, S.; Ishak, A. Three-dimensional flow and heat transfer of a nanofluid past a permeable stretching sheet with a convective boundary condition. *AIP Conf. Proc.* **2014**, *1614*, 906.
13. Hayat, T.; Muhammad, T.; Alsaedi, A.; Alhuthali, M.S. Magnetohydrodynamic three-dimensional flow of viscoelastic nanofluid in the presence of nonlinear thermal radiation. *J. Magn. Magn. Mater.* **2015**, *385*, 222–229. [CrossRef]

14. Hayat, T.; Aziz, A.; Muhammad, T.; Alsaedi, A. On magnetohydrodynamic three-dimensional flow of nanofluid over a convectively heated nonlinear stretching surface. *Int. J. Heat Mass Transf.* **2016**, *100*, 566–572. [CrossRef]
15. Muhammad, T.; Alsaedi, A.; Hayat, T.; Shehzad, S.A. A revised model for Darcy-Forchheimer three-dimensional flow of nanofluid subject to convective boundary condition. *Results Phys.* **2017**, *7*, 2791–2797. [CrossRef]
16. Hayat, T.; Muhammad, T.; Shehzad, S.A.; Alsaedi, A. An analytical solution for magnetohydrodynamic Oldroyd-B nanofluid flow induced by a stretching sheet with heat generation/absorption. *Int. J. Therm. Sci.* **2017**, *111*, 274–288. [CrossRef]
17. Muhammad, T.; Alsaedi, A.; Shehzad, S.A.; Hayat, T. A revised model for Darcy-Forchheimer flow of Maxwell nanofluid subject to convective boundary condition. *Chin. J. Phys.* **2017**, *55*, 963–976. [CrossRef]
18. Selimefendigil, F.; Oztop, H.F. Mixed convection of nanofluids in a three dimensional cavity with two adiabatic inner rotating cylinders. *Int. J. Heat Mass Transf.* **2018**, *117*, 331–343. [CrossRef]
19. Sheikholeslami, M.; Hayat, T.; Muhammad, T.; Alsaedi, A. MHD forced convection flow of nanofluid in a porous cavity with hot elliptic obstacle by means of Lattice Boltzmann method. *Int. J. Mech. Sci.* **2018**, *135*, 532–540. [CrossRef]
20. Mahanthesh, B.; Gireesha, B.J.; Animasaun, I.L.; Shashikumar, T.M.a.S. MHD flow of SWCNT and MWCNT nanoliquids past a rotating stretchable disk with thermal and exponential space dependent heat source. *Phys. Scr.* **2019**, *94*, 085214. [CrossRef]
21. Hu, Z.; Lu, W.; Thouless, M.D. Slip and wear at a corner with Coulomb friction and an interfacial strength. *Wear* **2015**, *338*, 242–251. [CrossRef]
22. Hu, Z.; Lu, W.; Thouless, M.D.; Barber, J.R. Effect of plastic deformation on the evolution of wear and local stress fields in fretting. *Int. J. Solids Struct.* **2016**, *82*, 1–8. [CrossRef]
23. Wang, H.; Hu, Z.; Lu, W.; Thouless, M.D. The effect of coupled wear and creep during grid-to-rod fretting. *Nucl. Eng. Des.* **2017**, *318*, 163–173. [CrossRef]
24. von Karman, T. Uberlaminare und turbulente Reibung. *Z. Angew. Math. Mech. ZAMM* **1921**, *1*, 233–252. [CrossRef]
25. Cochran, W.G. The flow due to a rotating disk. *Math. Proc. Camb. Philos. Soc.* **1934**, *30*, 365–375. [CrossRef]
26. Millsaps, K.; Pohlhausen, K. Heat transfer by laminar flow from a rotating disk. *J. Aeronaut. Sci.* **1952**, *19*, 120–126. [CrossRef]
27. Ackroyd, J.A.D. On the steady flow produced by a rotating disk with either surface suction or injection. *J. Eng. Math.* **1978**, *12*, 207–220. [CrossRef]
28. Miclavcic, M.; Wang, C.Y. The flow due to a rough rotating disk. *Z. Angew. Math. Phys.* **2004**, *54*, 1–12. [CrossRef]
29. Attia, H.A. Steady flow over a rotating disk in porous medium with heat transfer. *Nonlinear Anal.-Model. Control* **2009**, *14*, 21–26
30. Turkyilmazoglu, M.; Senel, P. Heat and mass transfer of the flow due to a rotating rough and porous disk. *Int. J. Therm. Sci.* **2013**, *63*, 146–158. [CrossRef]
31. Rashidi, M.M.; Kavyani, N.; Abelman, S. Investigation of entropy generation in MHD and slip flow over a rotating porous disk with variable properties. *Int. J. Heat Mass Transf.* **2014**, *70*, 892–917. [CrossRef]
32. Hatami, M.; Sheikholeslami, M.; Ganji, D.D. Laminar flow and heat transfer of nanofluid between contracting and rotating disks by least square method. *Powder Technol.* **2014**, *253*, 769–779. [CrossRef]
33. Mustafa, M.; Khan, J.A.; Hayat, T.; Alsaedi, A. On Bodewadt flow and heat transfer of nanofluids over a stretching stationary disk. *J. Mol. Liq.* **2015**, *211*, 119–125. [CrossRef]
34. Sheikholeslami, M.; Hatami, M.; Ganji, D.D. Numerical investigation of nanofluid spraying on an inclined rotating disk for cooling process. *J. Mol. Liq.* **2015**, *211*, 577–583. [CrossRef]
35. Hayat, T.; Muhammad, T.; Shehzad, S.A.; Alsaedi, A. On magnetohydrodynamic flow of nanofluid due to a rotating disk with slip effect: A numerical study. *Comput. Methods Appl. Mech. Eng.* **2017**, *315*, 467–477. [CrossRef]
36. Mustafa, M. MHD nanofluid flow over a rotating disk with partial slip effects: Buongiorno model. *Int. J. Heat Mass Transf.* **2017**, *108*, 1910–1916. [CrossRef]
37. Hayat, T.; Haider, F.; Muhammad, T.; Alsaedi, A. On Darcy-Forchheimer flow of carbon nanotubes due to a rotating disk. *Int. J. Heat Mass Transf.* **2017**, *112*, 248–254. [CrossRef]

38. Pop, I.; Soundalgekar, V.M. The Hall effect on an unsteady flow due to a rotating infinite disc. *Nucl. Eng. Des.* **1977**, *44*, 309–314. [CrossRef]
39. Bachok, N.; Ishak, A.; Pop, I. Flow and heat transfer over a rotating porous disk in a nanofluid. *Phys. B Condens. Matter* **2011**, *406*, 1767–1772. [CrossRef]
40. Lok, Y.Y.; Merkin, J.H.; Pop, I. Axisymmetric rotational stagnation-point flow impinging on a permeable stretching/shrinking rotating disk. *Eur. J. Mech. B/Fluids* **2018**, *72*, 275–292. [CrossRef]

MDPI
St. Alban-Anlage 66
4052 Basel
Switzerland
Tel. +41 61 683 77 34
Fax +41 61 302 89 18
www.mdpi.com

Symmetry Editorial Office
E-mail: symmetry@mdpi.com
www.mdpi.com/journal/symmetry

www.ingramcontent.com/pod-product-compliance
Lightning Source LLC
LaVergne TN
LVHW070846170726
843515LV00005B/588

9783036574905